T0275955

SpringerBriefs in Microbiology

More information about this series at http://www.springer.com/series/8911

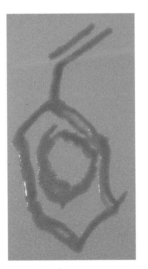

Gordonia sp. CWB8 isolated with styrene as sole source of carbon and energy

Dirk Tischler

Microbial Styrene Degradation

 Springer

Dirk Tischler
Institute of Biosciences, Environmental
  Microbiology
TU Bergakademie Freiberg
Freiberg
Germany

ISSN 2191-5385   ISSN 2191-5393   (electronic)
SpringerBriefs in Microbiology
ISBN 978-3-319-24860-8   ISBN 978-3-319-24862-2   (eBook)
DOI 10.1007/978-3-319-24862-2

Library of Congress Control Number: 2015950462

Springer Cham Heidelberg New York Dordrecht London

Printed on acid-free paper

Springer International Publishing AG Switzerland is part of Springer Science+Business Media
(www.springer.com)

Preface

The introduction of styrene into the environment by anthropogenic activity initiated numerous studies on its bioavailability and degradation by microorganisms already during the 1970s. Besides several studies with respect to eco-toxicity and environmental fate, also its effect on humans after exposure was investigated. It has been often reported that styrene itself and metabolites have serious impacts on the living world.

The historical time line started with the first description of styrene from a natural resource called Styrax in 1831. After heating, the plant material styrene was extracted and identified as the substance causing a distinct odor. Thus, its name was given based on the plant source. Styrene became a major chemical for diverse industries as, for example, in polymer-production. Later, the environmental pollution by styrene was investigated and from a microbiological point of view the degradation potential and pathways were described in detail. Single enzymes were also investigated. Thus, it was possible to solve mechanisms and structures as well as demonstrate biotechnological applications. Cascades of some enzymes or even complete pathways were either applied in bioremediation or production of fine chemicals. Respectively, over the years the field of microbial styrene conversion and related topics became a diverse and large area of research. Furthermore, styrene is one of the most produced and processed chemicals worldwide. Thus, its impact on the environment and ecosystems is still important!

In the group of Prof. Michael Schlömann the bacterial degradation of aromatic compounds and involved enzymes were studied extensively. While doing so also an enzyme possibly participating in styrene degradation was determined by Dirk Eulberg and Silvia Lakner. This monooxgyenase, later named StyA2B, was the first self-sufficient one-component styrene monooxygenase that was able to initially activate styrene for microbial breakdown. However, nothing was known about it and its biochemical role or even biotechnological application. That was the initiation of my Ph.D. project which aimed at the extensive characterization of StyA2B and related monooxygenases. During these studies several aspects of microbial styrene mineralization were investigated, enzymes studied, and biotechnological applications initiated. Respectively, several novelties were uncovered and published in order to contribute to the field of microbial styrene degradation.

The objective of this book is to present an overview of styrene degrading organisms and their repertoire of enzymes involved in metabolization of styrene under anaerobic and aerobic conditions. Therefore, a description of the physicochemical properties of styrene is given. Toxicological aspects are discussed. Microorganisms and their styrene degradation routes are outlined with respect to metabolites, genetic background, and regulatory machinery. Respectively, most important enzymes involved are highlighted. Furthermore, biotechnological applications that were reported or even patented are presented. The overview allows to determine the impact on various fields as microbiology, biochemistry, and biotechnology as well as poses still uncovered questions! Thus the book presents a general view of the current state of microbial styrene metabolism and initiates further work!

Funding is a general issue, especially for young scientists, and at this point I want to mention and thank my sources and promoters: Deutsche Bundesstiftung Umwelt, Fulbright, Saxonian Government, European Social Fund, American Society for Microbiology, collaborating institutes at Wageningen University, University Leipzig and San Francisco State University, Freunde und Förderer der TU Bergakademie Freiberg, and of course the Institute of Biosciences at TU Bergakademie Freiberg.

The past years we worked together as friends, students, and colleagues on several aspects of microbial styrene metabolism and respective enzymes. Thus, it was possible to collect and publish novel data as well as compile the literature from many years. Finally, it was possible to generate this book summarizing the state of the art on styrene metabolism by microorganisms including genetic, enzymatic, and technological reports. Herewith I want to say **THANK YOU** and acknowledge Michael Schlömann, Willem van Berkel, George Gassner, Andreas Schmid, Uwe Bornscheuer, Andreas Liese, Hedda Schlegel-Starmann, Stefan Kaschabek, Adrie Westphal, Janosch Gröning, Rene Kermer, Michel Oelschlägel, Anika Riedel, Thomas Heine, Stefania Montersino, Eliot Morrison, Ringo Schwabe, Paula Zwicker, Juliane Zimmerling, Catleen Conrad, and of course all other lab mates! Some of them supported to some extent the development of the book and provided content of chapters and will be therefore mentioned as co-authors in the acknowledgment.

As we all know, to achieve novelty and success depends not just on a good working situation, but to a large extent, also on the family behind. Therefore, I finally thank my parents, sister, wife, and son for never-ending support!

Finally, I thank the Springer publishing team, especially Jutta Lindenborn, for help throughout the writing and editing process. The overall support made it possible to create this overview of microbial styrene degradation and involved enzymes!

Freiberg   Dirk Tischler

Acknowledgments

At this point I thank again some colleagues for providing material presented in the chapters. Michel Oelschlägel (Chaps. 2–5) worked on the styrene oxide isomerase, phenylacetic acid production, and catabolism. Together with Anika Riedel (Chaps. 4 and 5) and Thomas Heine (Chaps. 4 and 5) we worked on the styrene monooxygenase. And Juliane Zimmerling (Chap. 4) provided input on the phenylacetaldehyde dehydrogenase. I also thank Judith S. Tischler for carefully proofreading the manuscript.

Contents

Abbreviations

ADP	Adenosine diphosphate
ATP	Adenosine triphosphate
ATPase	Adenosinetriphosphatase
BNAH	1-benzyl-1,4-dihydronicotinamide
BTEX	Benzene, toluene, ethylbenzene, and xylenes
CiP	Fungal peroxidase
CoA	Coenzyme A
CYP	Cytochrome P450 monooxygenase
EH	Epoxide hydrolase
$FAD_{ox/red}$	Oxidized (ox)/reduced (red) flavinadenine dinucleotide
feaB, FeaB	Phenylacetaldehyde dehydrogenase gene, or protein
FMO	Flavin-containing monooxygenase
HK	Histidine kinase
LPO	Lactoperoxidase
mEH	Microsomal epoxide hydrolase
$NAD(P)^+/NAD(P)H$	Oxidized/Reduced form of Nicotinamide adenine dinucleotide (phosphate)
NDDH	Naphthalene dihydrodiol dehydrogenase
p-HO styrene	4-hydroxystyrene
PaaABCDE	Ring 1,2-phenylacetyl-CoA epoxidase
PaaF	2,3-dehydroadipyl-CoA hydratase
PaaG	Ring 1,2-epoxyphenylacetyl-CoA isomerase
PaaH	3-hydroxyadipyl-CoA dehydrogenase
PaaJ	3-oxoadipyl-CoA/3-oxo-5,6-dehydrosuberyl-CoA thiolase
PaaK	Phenylacetate-CoA ligase
PaaZ	Oxepin-CoA hydrolase/3-oxo-5,6-dehydrosuberyl-CoA semialdehyde dehydrogenase
PAD	Phenylacetaldehyde dehydrogenase
PAR	Phenylacetaldehyde reductase
PTDH	Phosphite dehydrogenase

RR	Response regulator
SDD	Styrene 2,3-dihydrodiol dehydrogenase
SDO	Styrene 2,3-dioxygenase
SMO	Styrene monooxygenase
SOI	Styrene oxide isomerase
SOR	Styrene oxide reductase
sp.	A single species of a genus
spp.	Many species of a genus
sty	Styrene
styABCD	Styrene catabolic genes
StyABCD	Proteins obtained from *styABCD*-expression (see SMO, SOI, PAD)
styE	Encodes a cellular transporter for styrene
stySR	Regulatory elements of the *sty*-operon
VC12DO	Vinylcatechol 1,2-dioxygenase
VC23DO	Vinylcatechol 2,3-dioxygenase
vol	Volume

Chapter 1
Styrene: An Introduction

Abstract Styrene also known as cinnamene, ethenylbenzene, phenylethene, phenylethylene, and vinylbenzene is the simplest alkenylbenzene and one of the most produced and processed monoaromatic compounds worldwide. It is a colorless, viscous liquid which was first described in 1830s as a product from heating styrax oil (storax). This natural balsam consists mainly of cinnamates (45 %) and some vanillin (3 %). However, it has a distinct odor caused by traces of styrene. The latter has in its pure form a pungent or penetrating sweetish odor and was named according the respective plant material. Besides such natural sources, the alkenylbenzene is released in considerable amounts into the environment, mainly into atmosphere, due to anthropogenic activity. And in combination with its high chemical reactivity and environmental presence, one can expect styrene-catabolic routes to be widespread among microorganisms. Indeed, it was several times shown that styrene can be rapidly degraded under aerobic as well as anaerobic conditions. Therewith, it can be classified as readily biodegradable. It does not accumulate in soil or aquifer most likely due to its volatility and susceptibility to photooxidation. The chapter summarizes the properties of styrene and various aspects of its importance for human and nature in general. Additionally, it provides a broad view on what is presented and discussed in following chapters.

Keywords Styrene · Ecotoxicity · Environmental fate · Natural occurrence

1.1 Distribution of Styrene

Styrene occurs naturally in lignite tar and coal tar. Furthermore, it is a component of volatile and oily substances from plants and food (fruits, vegetables, nuts, meat) and can be produced by microorganisms as well (Khaksar and

© The Author(s) 2015
D. Tischler, *Microbial Styrene Degradation*,
SpringerBriefs in Microbiology, DOI 10.1007/978-3-319-24862-2_1

Ghazi-Khansari 2009; Lafeuille et al. 2009; Sielicki et al. 1978; Smith 1994; Warhurst and Fewson 1994). For example, *Penicillium* species are able to convert *p*-hydroxycinnamic acid and *p*-coumaric into styrene (Fu and Alexander 1992; Spinnler et al. 1992). Furthermore, styrene is released from decaying plant material (Shirai and Hisatsuka 1979). The natural occurrence of styrene indicates the presence of ubiquitous metabolic styrene pathways. Especially, the degradative routes are described in Chap. 2 and their molecular genetic background in Chap. 3, respectively.

The industrial synthesis of styrene proceeds mainly via a direct catalytical dehydration of ethylbenzene (Westblad et al. 2002). Since 1940s, the demand of styrene is increasing due to the synthesis of polyesters and plastics (Khaksar and Ghazi-Khansari 2009). Exemplarily, 3.6 million tons of styrene were produced in 1990 just in the United States of America (Fu and Alexander 1992). In 1996 and 1999, the world production was estimated to about 15 and 23 million tons per year, respectively. The latter values fit to the estimated global styrene demand of about 24 million tons solely for the year 2004 (CMAI 2005). In 2008, the global production was estimated to about 30 million tons and about 87 % were consumed by various industries (EU Commission DG ENV 2011). Especially, the polystyrene production can be accounted for a turnover of about 11 million tons of styrene.

Most of the styrene produced is applied to produce polystyrene (CMAI 2005; EU Commission DG ENV 2011; Tischler and Kaschabek 2012; van Agteren et al. 1998; Westblad et al. 2002). Besides this polymer other products are generated, as for example styrene-butadiene rubber, styrene acrylonitrile polymer, acrylonitrilie butadiene styrene resins, and diverse polyester resins. Further, the monomer is used as organic solvent in various processes.

The high industrial demand of styrene leads along production, processing, and transportation to its exposure and thus to serious environmental issues. In addition, styrene is released into environment by motor vehicle exhaust, styrene-polymer production and disposal, tobacco smoke, and combustion or pyrolysis processes. A further source is the migration of monomer styrene from polymers used as packing material in food industry (Khaksar and Ghazi-Khansaris 2009). These processes or sources of monomer styrene mainly result in atmospheric contamination, but its release in soil, aquifer, and open waters was also reported (Alexander 1990; CMAI 2005; European Union Risk Assessment Report Styrene 2002; Fu and Alexander 1992; US Inventory of Toxic Compounds 2001; van Agteren et al. 1998).

1.2 Physicochemical Properties of Styrene

Styrene represents the simplest alkenylbenzene and is an unsaturated aromatic compound (REAXYS database no. 1071236). Selected physicochemical properties are summarized in Table 1.1. The pure monomer is a colorless, oily liquid with a pungent odor which is comparable of the benzene odor. It is strongly

Table 1.1 Selected properties of styrene

Name	Styrene
Structure	
CAS no.	100-42-5
Molecular formula	C_8H_8
Molecular weight (g mol^{-1})	104.15
Heat of polymerization (kJ mol^{-1})	−69.8
Melting point or Freezing point (°C)	−30.63
Boiling point (°C)	145.15
Flash point (closed up; °C)	31
Density at 20 °C (g ml^{-1})	0.905
Vapor density relative to air	3.55
Vapor pressure at 20 and 30 °C (mm Hg)	5 and 9.5
Water solubility at 15, 20, and 30 °C (mg l^{-1})	280, 300, and 400
Henry partition coefficient (air/water)	0.21
Octanol/water partition coefficient (K_{ow})	3.02 (log value)
LD$_{50}$ rat (oral application; mg kg^{-1})	1000
LC$_{50}$ rat (inhalation; 4 h; mg m^3)	12,000

refractive. Styrene is flammable and miscible with most organic solvents, but only slightly soluble in water. The monomer tends to spontaneous polymerization due to the high reactivity of the exocyclic double bond. Therefore, the monomer needs to be stabilized which is often achieved by addition of 4-*tert*-butylcatechol. Even at room temperature it reacts to polystyrene under light influence. Without a polymerization inhibitor, it also starts to polymerize by heating in the presence of molecular oxygen. At higher temperatures, the exothermic reaction gets self-accelerated and results in an explosive mixture containing peroxy-structures.

1.3 Toxicology and Environmental Fate of Styrene

Extensive industrial production and application as well as other processes mentioned above (Sect. 1.2) lead to a substantial styrene entry into environment. Anthropogenic styrene release was estimated for the European Union to about 23,500 tons annually (European Union Risk Assessment Report Styrene 2002). And 96 % of it is supposed to be emitted directly into atmosphere. For a single country as the Netherlands with styrene-related industry, about 1500 tons gaseous emissions were reported (van Agteren et al. 1998). And for those reasons, environment and humans are faced to styrene exposure which caused numerous toxicological studies (Gibbs and Mulligan 1997; Khaksar and Ghazi-Khansaris 2009; Rueff et al. 2009).

Due to volatility styrene uptake occurs mainly via lungs and to some extent through the skin or by eating (Khaksar and Ghazi-Khansaris 2009; Tischler and Kaschabek 2012; van Agteren et al. 1998). It irritates exposed organs as eyes and membranes of mouth and nose. Once in a body styrene can accumulate in lipid deposits or it is biotransformed by hepatic microsomal cytochrome P450 monooxygenases, microsomal epoxide hydrolases (mEH), and dehydrogenases yielding styrene oxide, phenylethanediol, mandelic acid, and phenylglyoxylic acid, respectively (Hartmans 1995; Tischler and Kaschabek 2012). The latter two acids are used as biomarkers for styrene exposure since their urinary detection correlates with styrene uptake (Guillemin and Berode 1988). Neurotoxic effects to mammals can be observed after contact with high concentrations of styrene and may reflect the interaction of styrene with nerve cell membranes (Bond 1989). Styrene causes damage to both peripheral and central nervous system after chronic exposure (Marczynski et al. 2000; Murata et al. 1991; Rebert and Hall 1994). Further, it was found to have immunomodulatory effects on humans after inhalation (Tulinska et al. 2000). Thus, styrene itself can be classified as genotoxic and priority toxic compound. Furthermore, the monomer styrene was classified as carcinogenic to humans (group 2B) by the International Agency for Research on Cancer (IARC). The potential of carcinogenicity is most likely due to the styrene metabolism of mammals which form styrene oxide. This epoxide is highly reactive and able to covalently modify DNA molecules causing respective mutagenic effects. And it was found that styrene oxide itself has drastic effects on the central nervous system (Marczynski et al. 2000; Mutti 1988; Mutti et al. 1988).

1.4 Motivations to Study Styrene from a Bioperspective

With the rapidly growing industry and the consumer-affected society, effects of mankind on the environment got stronger and more obvious! It was realized that they have an impact on life. Especially, the anthropogenic and even uncontrolled release of toxic and recalcitrant hydrocarbons into the environment induced several concerns regarding the question: "How dealing with such contaminants or contaminated sites?" Understanding the bioavailability, persistence, any degradative potential, and the ecotoxicity of such substances came into the focus of scientists. The microbiology community contributed in solving these issues with numerous studies revealing, for example, many degradative pathways and microbial powerhouses. Interestingly, many of such pathways for hydrocarbons dumped by man came later in focus for biotechnological applications. Especially, because aliphatic and aromatic compounds need to be activated and microorganisms achieve it often by a selective oxygenation process. And such regio- and stereoselective oxygenations are of high interest for many chemical syntheses.

The microbial degradation of styrene and its environmental fate combines all attributes mentioned. Styrene is produced in huge quantities by industry, released due to industrial spillage into environment, and determined to be toxic for many organisms. However, it is also readily biodegradable and intermediates which occur during its degradation as well as the styrene degrading microorganisms are of biotechnological relevance. Therefore, styrene can be seen as a model substance studying microbial degradation of aromatic compounds and uncovering novel enzymes for regio- and stereoselective biotransformations. The understanding of the styrene biodegradation, involved enzymes, and biotechnological aspects will be highlighted in the following chapters.

References

Alexander M (1990) The environmental fate of styrene. SIRC Rev 1:33–42
Bond JA (1989) Review of the toxicology of styrene. Crit Rev Toxicol 19:227–249
CMAI (2005) Chemical Market Associates, Inc. 2006 World Styrene Analysis (2005) Houston. Texas, USA
EU Commission DG ENV (2011) Plastic waste in the environment. A report: http://ec.europa.eu/environment/waste/studies/pdf/plastics.pdf
European Union Risk Assessment Report Styrene (2002) Styrene, Part I—Environment. European Communities, Luxembourg
Fu MH, Alexander M (1992) Biodegradation of styrene in samples of natural environments. Environ Sci Technol 26:1540–1544
Gibbs BF, Mulligan CN (1997) Styrene toxicity: an ecotoxicological assessment. Ecotoxicol Environ Saf 38:181–194
Guillemin MP, Berode M (1988) Biological monitoring of styrene: a review. Am Ind Hyg Assoc J 49:497–505
Hartmans S (1995) Microbial degradation of styrene. In: Biotransformations: microbiological degradation of health risk compounds. Elsevier Science 32:227–238
Khaksar M-R, Ghazi-Khansari M (2009) Determination of migration monomer styrene from GPPS (general purpose polystyrene) and HIPS (high impact polystyrene) cups to hot drinks. Toxicol Mech Method 19:257–261
Lafeuille J-L, Buniak M-L, Vioujas M-C, Lefevre S (2009) Natural formation of styrene by cinnamon mold flora. J Food Sci 74:276–283
Marczynski B, Peel M, Baur X (2000) New aspects in genotoxic risk assessment of styrene exposure—a working hypothesis. Med Hypotheses 54:619–623
Murata K, Araki S, Yokoyama K (1991) Assessment of the peripheral, central, and autonomic nervous system function in styrene workers. Am J Ind Med 20:775–784
Mutti A (1988) Styrene exposure and serum prolactin. J Occup Med 30:481–482
Mutti A, Falzoi M, Romanelli A, Bocchi MC, Ferroni C, Franchini I (1988) Brain dopamine as a target for solvent toxicity: effects of some monocyclic aromatic hydrocarbons. Toxicology 49:77–82
REAXYS database including Beilstein and Gmelin (2014) Styrene: no. 1071236. Copyright 2014 Reed Elsevier Properties SA
Rebert CS, Hall TA (1994) The neuroepidemiology of styrene: a critical review of representative literature. Crit Rev Toxicol 24:S57–S106
Rueff J, Teixeira JP, Santos LS, Gaspar JF (2009) Genetic effects and biotoxicity monitoring of occupational styrene exposure. Clin Chim Acta 399:8–23

Shirai K, Hisatsuka K (1979) Production of β-phenethyl alcohol from styrene by *Pseudomonas* 305-STR-1-4. Agric Biol Chem 43:1399–1406

Sielicki M, Focht DD, Martin JP (1978) Microbial transformations of styrene and [14C] styrene in soil and enrichment cultures. Appl Environ Microbiol 35:124–128

Smith MR (1994) The physiology of aromatic hydrocarbons degrading bacteria. Biochemistry of microbial degradation. Kluwer, Dordrecht, pp 347–378

Spinnler HE, Grosjean O, Bouvier I (1992) Effect of parameters on the production of styrene (vinyl benzene) and 1-octene-3-ol by *Penicillium caseicolum*. J Dairy Res 59:533–541

Tischler D, Kaschabek SR (2012) Microbial degradation of xenobiotics. In: Singh SN (ed). Springer, Berlin, pp 67–99

Tulinska J, Dusinaska M, Jahnova E, Liskova A, Kuricova M, Vodicka P, Vodickova L, Sulcova M, Fuortes L (2000) Changes in cellular immunity among workers occupationally exposed to styrene in a plastics lamination plant. Am J Ind Med 38:576–583

US Inventory of Toxic Compounds (2001) TRI92. Toxics release inventory public data. Office of Pollution Prevention and Toxics, US EPA, Washington, DC, 94. Available at http://www.epa. gov

van Agteren MH, Keuning S, Janssen DB (1998) Handbook on biodegradation and biological treatment of hazardous organic compounds. Kluwer, Dordrecht, pp 235–242

Warhurst AM, Fewson CA (1994) A review. Microbial metabolism and biotransformation of styrene. J Appl Bacteriol 77:597–606

Westblad C, Levindis YA, Richter H, Howard JB, Carlson J (2002) A study on toxic emissions from batch combustion of styrene. Chemosphere 49:395–412

Chapter 2
Pathways for the Degradation of Styrene

Abstract The monomer styrene can be degraded by various microorganisms under aerobic and anaerobic conditions. Therefore, several peripheral pathways are employed yielding few central intermediates as 3-vinylcatechol, phenylacetic acid, benzoic acid, or 2-ethylhexanol. However, the anaerobic breakdown of styrene is less extensively described compared to the aerobic metabolization, and for the latter mainly Pseudomonads and Actinobacteria have been studied. There is only one styrene-specific pathway, designated side-chain oxidation, reported so far, while all other routes can be considered as unspecific. Thus microorganisms possessing pathways for toluene and biphenyl via direct ring cleavage, for example, can breakdown styrene as well. Besides these degradation capabilities, the partial metabolic activity of higher organisms is mentioned which often yields marker compounds, mandelic acid and phenylglyoxylic acid.

Keywords Aromatic degradation · Xenobiotic · Side-chain oxygenation · *Ortho*- and *meta*-cleavage · Oxygenase · Phenylacetic acid

2.1 Styrene-Degrading Microorganisms

The distribution of styrene among various environments due to a natural occurrence (Warhurst and Fewson 1994; Shirai and Hisatsuka 1979) or more importantly by industrial spillage makes the compound to a potential carbon source for microorganisms. For that, either unspecific routes may be used or even specific ones can evolve and allow microorganisms to use styrene as sole source of carbon and energy. Respective metabolisms for styrene have been found among several prokaryotic and eukaryotic organisms (Table 2.1). However, in the first studies regarding the bioavailability and mineralization of styrene mainly mixed cultures have been

© The Author(s) 2015
D. Tischler, *Microbial Styrene Degradation*,
SpringerBriefs in Microbiology, DOI 10.1007/978-3-319-24862-2_2

Table 2.1 Microorganisms capable to mineralize styrene

Class	Genus	References
Bacteria		
Actinobacteria	*Brevibacterium*	Hou et al. (1983)
	Corynebacterium	Itoh et al. (1996)
	Gordonia	Alexandrino et al. (2001), Oelschlägel et al. (2014a, b)
	Mycobacterium	Burback and Perry (1993)
	Nocardia	Furuhashi et al. (1986), Hartmans et al. (1990)
	Rhodococcus	Hartmans et al. (1990), Jung and Park (2005), Oelschlägel et al. (2012, 2014a, b), Patrauchan et al. (2008), Tischler et al. (2009), Toda and Itoh (2012), Warhurst et al. (1994a), Zilli et al. (2003)
	Streptomyces	Przybulewska et al. (2006)
	Tsukamurella	Arnold et al. (1997)
Bacilli	*Bacillus*	Przybulewska et al. (2006)
Clostridia	*Clostridium*	Grbić-Galić et al. (1990)
α-Proteobacteria	*Methylosinus*	Higgins et al. (1979)
	Sphingobium	Oelschlägel et al. (2014b)
	Sphingomonas	Arnold et al. (1997)
	Sphingopyxis	Oelschlägel et al. (2014b)
	Xanthobacter	Hartmans et al. (1989, 1990)
β-Proteobacteria	*Nitrosomonas*	Keener and Arp (1994)
γ-Proteobacteria	*Enterobacter*	Grbić-Galić et al. (1990)
	Methylococcus	Colby et al. (1977)
	Pseudomonas	Alexandrino et al. (2001), Baggi et al. (1983), Beltrametti et al. (1997), Bestetti et al. (1984), Gąszczak et al. (2012), Ikura et al. (1997), Kim et al. (2005), Lin et al. (2010), Marconi et al. (1996), O'Conner et al. (1995), Oelschlägel et al. (2014a), Panke et al. (1998), Park et al. (2006), Rustemov et al. (1992), Shirai and Hisatsuka (1979), Utkin et al. (1991), Velasco et al. (1998)
	Xanthomonas	Arnold et al. (1997)
Sphingobacteria	*Sphinogobacterium*	Przybulewska et al. (2006)
Fungi		
Agaricomycetes	*Bjerkandera*	Braun-Lüllemann et al. (1997)
	Phanerochaete	Braun-Lüllemann et al. (1997)
	Pleurotus	Braun-Lüllemann et al. (1997)
	Trametes	Braun-Lüllemann et al. (1997)
Dothideomycetes	*Cladosporium*	Weber (1995), Weber et al. (1995)
	Caldariomyces	Geigert et al. (1986)
Eurotiomycetes	*Aspergillus*	Paca et al. (2001)
	Exophiala	Cox et al. (1993, 1996)
	Penicillium	Cox (1995), de Jong et al. (1990), Paca et al. (2001)
Sordariomycetes	*Gliocladium*	Cox (1995)
	Sporothrix	Cox (1995), René et al. (2010)

investigated (Alexandrino et al. 2001; Araya et al. 2000; Arnold et al. 1997; Cox et al. 1993; Grbić-Galić et al. 1990; Lu et al. 2001; Weigner et al. 2001). Initially, Pseudomonads were investigated as well because of their well-known potential in breakdown of aromatic compounds (Baggi et al. 1983; Bestetti et al. 1984; Shirai and Hisatsuka 1979). Later, further pure strains as, for example, from the genus *Rhodococcus* were isolated and studied in detail (Table 2.1; see Sects. 2.2–2.4).

As already indicated, most of the studies describe the aerobic breakdown of styrene via phenylacetic acid by microorganisms (Tischler and Kaschabek 2012) in which especially the genus *Pseudomonas* was frequently investigated (O'Leary et al. 2002). However, in the meantime, it is reported that several Gram-negative and Gram-positive bacteria as well as fungi are able to mineralize styrene completely. Therefore, the monomer styrene seems to be readily bioavailable and may enter the cell via diffusion or by means of a specially evolved transport system (Mooney et al. 2006a; Nikodinovic-Runic et al. 2009). When investigated in dependence of the organism, styrene shows inhibitory effects on the growth behavior (Cox et al. 1993, 1997; Gąszczak et al. 2012; Tischler et al. 2009). However, *Pseudomonas* sp. E-93486 could utilize styrene as a carbon source even from media containing a styrene concentration of up to 90 g m^{-3} (Gąszczak et al. 2012). In comparison, the white-rot fungi *Pleurotus ostreatus* was able to degrade 37 g styrene m^{-3} within 48 h, whereas related fungi were not capable to utilize such styrene concentrations (Braun-Lüllemann et al. 1997).

Despite the presence or absence of molecular oxygen, two major strategies to activate and further degrade the monomer styrene have been identified (Mooney et al. 2006b; O'Leary et al. 2002). In both cases oxygen (molecular or from water) is used by microorganisms to activate styrene. This initial attack either targets the aromatic nucleus or the vinyl side-chain. The latter one seems to be styrene-specific since epoxidation (aerobic) or hydratation (anaerobic) of the vinyl side-chain has frequently been described for various microorganisms (Tischler and Kaschabek 2012). Whereas the initial mono- or dihydroxylation of the aromatic ring seems to be a product of unspecific pathways which are commonly responsible for the degradation of structurally related aromatic compounds such as biphenyl, ethylbenzene, or toluene (Cho et al. 2000; Patrauchan et al. 2008; Warhurst et al. 1994a). That is reasonable, since usually these routes can be induced by various compounds. And they can be passed through the same enzymatic machinery. This is possible due to relaxed substrate specificity of the respective enzymes and a similar reaction pattern (e.g. *meta*-cleavage of the aromatic nucleus). Analogous intermediates can often be observed for the substrates mentioned (Carmona et al. 2009; Patrauchan et al. 2008).

2.2 Aerobic Styrene Metabolism

The monomer styrene is either attacked at the vinyl side-chain or at its aromatic nucleus by means of mono- or dioxygenases, respectively. So far, most studies have revealed a route via side-chain oxidation which can therefore be supposed

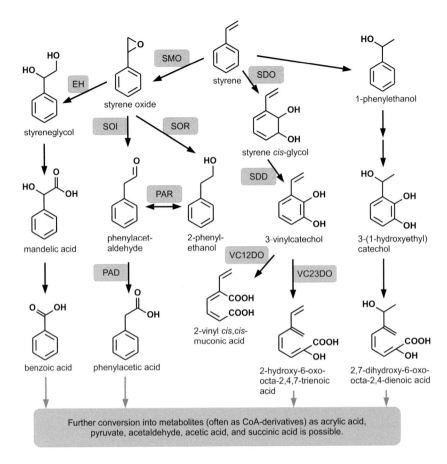

Fig. 2.1 Aerobic styrene metabolism

as the favored mechanism in nature (Beltrametti et al. 1997; Cox et al. 1996; Hartmans et al. 1990; Itoh et al. 1996; Oelschlägel et al. 2012, 2014b; Panke et al. 1998; Park et al. 2006; Toda and Itoh 2012; Velasco et al. 1998). In the following sections both pathways and, in addition, some modified routes are discussed (Fig. 2.1).

2.2.1 Vinyl Side-Chain Oxidation (Upper Styrene-Degrading Pathway)

The upper styrene pathway starts with an activation of the vinyl side-chain via oxidation in order to produce phenylacetic acid as a key product (Fig. 2.1). Numerous strains have been reported following this route, for example, *Corynebacterium* sp. AC-5 (Itoh et al. 1996), *Exophiala jeanselmei* (Cox et al. 1996), *Xanthobacter* sp. strain 124X (Hartmans et al. 1989), *Sphingopyxis fribergensis* Kp5.2 (Oelschlägel et al. 2014b, 2015) *Rhodococcus* sp. ST-5 (Toda and Itoh 2012), *Rhodococcus opacus* 1CP (Tischler et al. 2009; Oelschlägel et al. 2012), and several *Pseudomonas* strains (Kantz et al. 2005; Lin et al. 2010; Marconi et al. 1996; O'Connor et al. 1995, 1997; Panke et al. 1998; Park et al. 2006; Utkin et al. 1991; Velasco et al. 1998). In addition, some white-rot fungi also follow a side-chain oxidation route, but employing other enzymes and therefore yielding different intermediates (see Sect. 2.2.4) (Braun-Lüllemann et al. 1997).

First, styrene monooxygenases (SMO) use molecular oxygen to perform styrene epoxidation. Interestingly, bacteria have an FAD-dependent enzyme (reviewed by Ceccoli et al. 2014; Huijbers et al. 2014; Montersino et al. 2011), whereas fungal species contain a heme dependent cytochrome P450 monooxygenase (Cox et al. 1996). In the case of bacterial SMOs, the conversion is highly enantioselective and yields the respective *S*-enantiomer of styrene oxide. The epoxide formed gets further converted by styrene oxide isomerase (SOI) into phenylacetaldehyde while the chiral information introduced gets lost again. Typically, the SOI is a transmembrane protein and cofactor independent (Itoh et al. 1997; Oelschlägel et al. 2012). Some preference of the SOI for the *S*-enantiomer of styrene oxide has also been shown (Itoh et al. 1997; Miyamoto et al. 2007; Oelschlägel et al. 2012). Next, the NAD-consuming phenylacetaldehyde dehydrogenase (PAD) catalyzes the reaction of phenylacetaldehyde into phenylacetic acid. As side-reaction unspecific dehydrogenseases and/or reducases might convert phenylacetaldehyde into 2-phenylethanol and vise versa (Beltrametti et al. 1997; Marconi et al. 1996).

Finally, this peripheral route to transform styrene into phenylacetic acid can be seen as the only styrene-specific pathway identified so far (Mooney et al. 2006b; Tischler and Kaschabek 2012). That gets obvious from the following points: (i) the pathway is positively regulated by styrene (Velasco et al. 1998), (ii) it comprises enzymes which are highly specific for styrene(s) and corresponding metabolites, (iii) knockout mutants for key enzymes lost ability to grow on styrene as sole carbon source (Han et al. 2006), (iv) only styrene derivatives can be converted to respective acids but at poor rates (Oelschlägel et al. 2014a), and (v) neither such styrene derivatives nor other aromatic compounds serve as sole carbon sources and could not be degraded via this peripheral route which was exemplarily shown for a key enzyme of this pathway (Itoh et al. 1997) or the co-metabolic turnover of styrene derivatives (Oelschlägel et al. 2015).

2.2.2 Phenylacetic Acid Catabolism (Lower Styrene-Degrading Pathway)

The side-chain oxidation of styrene can be seen as a peripheral catabolic route yielding the central intermediate phenylacetic acid (Navarro-Llorens et al. 2005; Olivera et al. 1998; Teufel et al. 2010). Besides styrene, also phenylalanine,

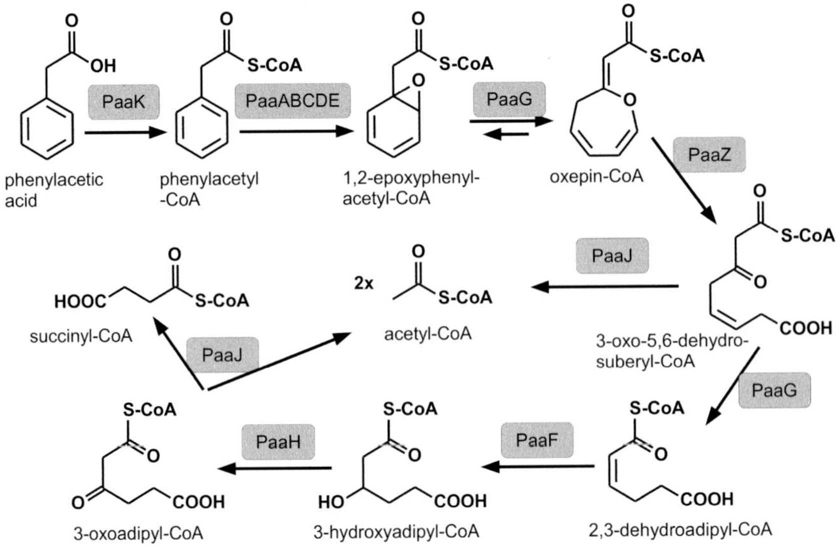

Fig. 2.2 Aerobic catabolism of phenylacetic acid

2-phenylethanol, phenylethylamine, *trans*-styrylacetic acid, phenylalkanoic acids, and tropic acid also lead to this intermediate. However, the upper and lower styrene-degrading pathways are connected via regulatory elements (Peso-Santos et al. 2006) and therefore the phenylacetic acid catabolism will be outlined herein. Further, it should be mentioned that this pathway is the only one elucidated for aerobic phenylacetic acid so far, while present in about 16 % of all genome-sequenced microorganisms (Teufel et al. 2010). The key enzymes and reactions (Fig. 2.2) are conserved but different types of regulators (PaaR and PaaX) have been described, respectively (Chen et al. 2012).

Phenylacetic acid can be actively imported by transporter proteins (PaaL, PaaP) (Peso-Santos et al. 2006) or produced from styrene as mentioned above. Then an ATP-consuming reaction of a phenylacetate-CoA ligase (PaaK) yields phenylacetyl-CoA. The thioester obtained gets activated at the aromatic nucleus to the 1,2-epoxide by means of a multicomponent epoxidase (PaaABCDE) (Teufel et al. 2010). Next, the ring 1,2-epoxyphenylacetyl-CoA isomerase (PaaG) isomerizes the reactive epoxide to an oxepin-CoA intermediate which undergoes a hydrolytic ring cleavage. During the latter step NADPH is formed by an oxepin-CoA hydrolase (PaaZ) and β-oxidation steps follow to yield acetyl-CoA and succinyl-CoA (by PaaJ, PaaG, PaaF, PaaH) (Navarro-Llorens et al. 2005; Teufel et al. 2010).

2.2.3 Direct Ring Cleavage of Styrene

A direct attack of the aromatic nucleus by mono- or dioxygenases is another possibility to activate styrene for a further breakdown (Fig. 2.1). However, a dioxygenation and a later *meta*-cleavage was found to be prominent among several microorganisms as, for example, *Rhodococcus rhodochrous* NCIMB 13259 (Warhurst et al. 1994a), *Pseudomonas* sp. Y2 (Utkin et al. 1991), X*antobacter* sp. strain 124X (Hartmans et al. 1989), *Pseudomonas putida* MST (Bestetti et al. 1989), and *Rhodococcus jostii* RHA1 (Patrauchan et al. 2008). Also the contemporary presence of side-chain oxygenation and direct ring cleavage in a single microorganism in an active form seems possible (Hartmans et al. 1989; Utkin et al. 1991).

Warhurst and coworkers (1994a) have described in detail a probable styrene-catabolic pathway via ring hydroxylation and *meta*-cleavage in detail. The sequential activity of styrene 2,3-dioxygenase (SDO) and styrene 2,3-dihydrodiol dehydrogenase (SDD) initially dihydroxylate styrene to styrene *cis*-glycol and then catalyze the rearomatization of this glycol to yield 3-vinylcatechol. Considering catechol-like compounds as the key intermediates, these two peripheral steps are similar to those which are involved in the degradation of benzene-, ethylbenzene-, and toluene, respectively (Mars et al. 1997; Smith 1990; Warhurst et al. 1994b). The 3-vinylcatechol obtained can then undergo *ortho*- or *meta*-cleavage. Both have been found in strain NCIMB 13259, in which, however, the activity of the vinylcatechol 1,2-dioxygenase (VC12DO) leads to the *ortho*-product 2-vinyl-*cis,cis*-muconic acid which represents a dead-end product (Warhurst et al. 1994a, b). Respectively, the *meta*-route including the vinylcatechol 2,3-dioxygenase (VC23DO) resembles the catabolically relevant one leading to 2-hydroxy-6-oxo-octa-2,4,7-trienoic acid. The latter product is further transformed into acrylic acid, acetaldehyde, and pyruvate. These observations are in congruence with results obtained from the degradation of various alkylcatechols (Knackmuss et al. 1976; Marín et al. 2010; Patrauchan et al. 2008).

After converting styrene to 1-phenylethanol, *Pseudomonas* sp. Y2 also showed the potential to directly attack the aromatic nucleus and to degrade the intermediate via a *meta*-pathway (Utkin et al. 1991). The enzymes involved were not studied, but a similar route as described above can be suggested leading to complete styrene mineralization via analogous intermediates.

However, it should be mentioned that this described styrene pathway is supposed to be rather unspecific (Tischler and Kaschabek 2012). Styrene itself seems not to be the original substrate, but funneled through the route due to a relaxed substrate specificity of involved enzymes. This is also shown by both *Rhodococcus* strains reported (Patrauchan et al. 2008; Warhurst et al. 1994a) which were able to utilize besides styrene benzene, ethylbenzene, toluene, and other aromatic compounds. In most cases a clear indication for a *meta*-cleavage route was found. Most striking is that strain RHA1 utilizes biphenyl as carbon source and even activates this compound faster than the others (Patrauchan et al. 2008). Genomic studies revealed the respective biphenyl operon and a knockout of the corresponding biphenyl dioxygenase abolished the growth of strain RHA1 on styrene. Respectively, the biphenyl catabolic operon enables this strain to co-metabolize styrene via a *meta*-cleavage pathway.

2.2.4 Alternative Routes to Mineralize or Transform Styrene

Marconi and coworkers (1996) have reported 2-phenylethanol as a minor metabolite obtained during styrene degradation activity of *Pseudomonas fluorescens* ST which usually performs side-chain oxygenation. Therefore, additional activities of styrene oxide reductase (SOR) and phenylacetaldehyde reductase (PAR) have been predicted by the above-mentioned study (Fig. 2.1). But, the compound 2-phenylethanol was determined to be a major metabolite for other bacteria, such as, *Pseudomonas* sp. 305-STR-1-4 (Shirai and Hisatsuka 1979), *Pseudomonas* sp. Y2 (Utkin et al. 1991), and *Xanthobacter* strain 124X (Hartmans et al. 1989). Furthermore, this styrene biotransformation might belong to an unspecific route which has been supposed to be the ethylbenzene pathway (Tischler and Kaschabek 2012).

Rhodococcus sp. ST-10 has been reported to harbor an incomplete styrene degradation route since no enzymatic SOI activity was determined which was suggested to be overcome by some other microbial activity from ecosystem or chemical isomerization (Toda and Itoh 2012). Respectively, no *styC*-gene has been identified for strain ST-10 yet. It was speculated that due to the chemical conversion of styrene oxide into the respective phenylacetaldehyde the strain might utilize styrene as carbon source. But, this assumption is rather unlikely since under conditions as occurring inside a cell, the epoxide isomerization probably does not occur (Han et al. 2006; Oelschlägel et al. 2012). Further, strain ST-10 yielded acetophenone and styreneglycol as dead-end metabolites (Toda and Itoh 2012). With respect to other possible routes (Fig. 2.1), the activity of SOR and PAR might in concert overcome the missing SOI activity. However, the latter activities have not been shown for strain ST-10 so far. Several other strains have been found to co-metabolize styrene yielding growth or enrichment of dead-end products, respectively. Thus, styrene is often epoxidized and not further catabolized as shown, for example, for *Nitrosomonas* or *Nocardia* (Furuhashi et al. 1986; Keener and Arp 1994). Otherwise *R. jostii* RHA1 was found to funnel styrene through its biphenyl pathway and can use so styrene as a sole carbon source (Patrauchan et al. 2008).

The difference between the bacterial and the fungal side-chain oxygenation route has already been mentioned (Sect. 2.2.1). In dependence of the microorganism studied, different enzymes for initial styrene epoxidation are employed. However, for a few white-rot fungi also the further catabolism differs significantly (Fig. 2.1) (Braun-Lüllemann et al. 1997). Herein, the epoxide formed is further converted to styreneglycol by means of an epoxide hydrolase (EH). The diol obtained is further oxidized to mandelic acid by a dehydrogenase and thereafter decarboxylated to yield benzoic acid. This route shows high similarity to the detoxification metabolism of human (Rueff et al. 2009). In addition, side products like 2-phenylethanol were also determined.

2.3 Anaerobic Routes for Styrene Degradation

Anaerobic mineralization of aromatic compounds is already known for some time, but the detailed investigations on pathways, genetics, enzymes, and the general capacity have just recently been reported and further work is still necessary (Carmona et al. 2009). Microbial degradation of aromatic compounds under anaerobic conditions is handicapped due to the delocalized electrons and therewith stabilized aromatic structure. Further, the absence of molecular oxygen does not allow hydroxylations and oxygenolytic ring cleavage reactions which are main processes in aerobic breakdown of such compounds. Respectively, reductive reactions are mainly recruited to attack and activate aromatic compounds. As for aerobic pathways described, a variety of peripheral routes lead to central intermediates of the anaerobic catabolism, for example, benzoyl-CoA and resorcinol. These pathways depend on the redox potential and electron acceptors applicable for the microorganism (Carmona et al. 2009).

However, not much information is available on the anaerobic metabolism of styrene. Pure cultures and microbial consortia obtained from anaerobic sludge were investigated for their potential in styrene mineralization (Grbić-Galić et al. 1990; Araya et al. 2000). The styrene breakdown was determined and respective metabolites were identified. Based on these intermediates pathways have been assigned (Fig. 2.3). Here two main metabolites have to be mentioned: phenylacetic acid and ethylphenol. Both intermediates can funnel via benzoyl-CoA into the central metabolism (Carmona et al. 2009). And interestingly, both consortia and

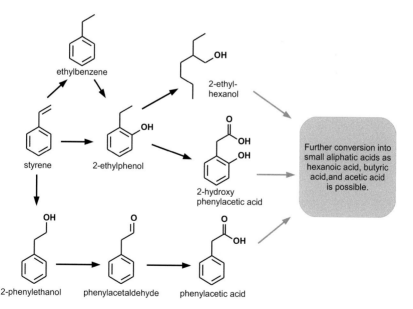

Fig. 2.3 Anaerobic styrene degradation routes based on metabolites determined (adapted from Grbić-Galić et al. 1990; Tischler and Kaschabek 2012)

pure cultures were able to almost completely mineralize styrene via those proposed routes. Carbon dioxide has been supposed to be the final product and only traces of aromatic or alicyclic intermediates remained in the media. But, respective gene clusters, regulatory elements, and enzymes have not been studied, so far, and therefore the anaerobic styrene breakdown remains uncovered.

Anaerobic consortia and *Enterobacter* have been found to degrade styrene via a peripheral route into phenylacetic acid (Grbić-Galić et al. 1990). It is similar to the aerobic styrene conversion via side-chain oxygenation. As the authors state, styrene is first transformed into 2-phenylethanol by addition of water catalyzed by a hydratase. Steps following are catalyzed by different dehydrogenses which transform 2-phenylethanol via phenylacetaldehyde into phenylacetic acid (Grbić-Galić et al. 1990). Alternatively, the aromatic ring of styrene can be directly attacked and the vinyl side-chain can be reduced yielding 2-ethylphenol. The latter compound can then be converted into 2-ethylhexanol by direct ring cleavage or oxidized into 2-hydroxyphenylacetic acid. However, both routes assigned for anaerobic styrene metabolism seem to be important and occurring contemporarily in the microbial consortia (Tischler and Kaschabek 2012).

2.4 Biotransformation of Substituted Styrene Compounds

Besides styrene itself, several microorganisms are capable to metabolize or even use substituted styrene compounds or oligomers as sole source of carbon and energy (Higashimura et al. 1983; Oelschlägel et al. 2014a, b; Omori et al. 1974; Tuschii et al. 1977; Warhurst et al. 1994a, b).

For example, α-methyl and β-methylstyrene can serve as substrates for bacteria (Omori et al. 1974; Warhurst et al. 1994a). Bestetti and coworkers (1989) were able to confirm observations earlier made on these biotransformations. *P. putida* strain MST was also isolated with α-methylstyrene as sole source of carbon and energy. Thus two different initial attacks on the styrene derivative have been postulated since 2-phenyl-2-propen-1-ol and 1,2-dihydroxy-3-isopropenyl-3-cyclohexene were determined from the culture broth (Fig. 2.4). No evidence for the side-chain oxidation and respective enzymes were determined as had been described earlier for another *Pseudomonas* strain (Baggi et al. 1983). But, the intermediates detected, which were hydroxylated at the aromatic ring, indicate the presence of a more unspecific degradation pathway (compare with Sect. 2.2.3). Furthermore, *R. rhodochrous* NCIMB 13259 is able to grow on α-methylstyrene as a sole carbon source but not on β-methylstyrene (Warhurst et al. 1994a). As for styrene, the strain might possess a direct ring cleavage route in order to activate and transform α-methylstyrene.

The biodegradation of halogenated styrenes has not been described so far. But, the bacterial conversion of, especially, chlorinated styrene's into various intermediates is possible (Hudlicky et al. 1993; Oelschlägel et al. 2014a). First, *P. putida* 39D and its toluene dioxygenase which is probably involved in the direct ring hydroxylation have been employed to convert styrene and chlorostyrenes into corresponding

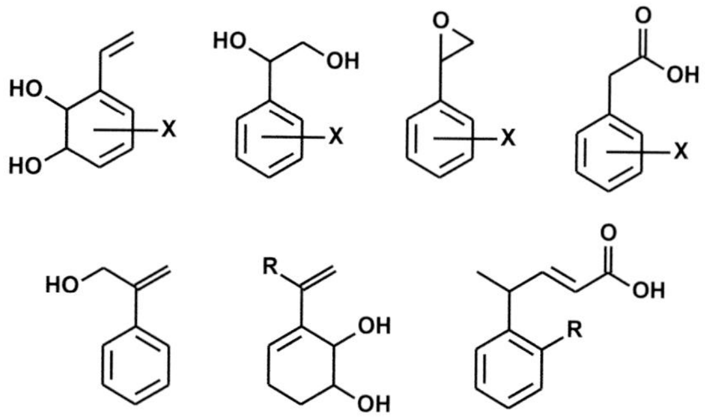

Fig. 2.4 Metabolites determined from biotransformations of styrene, its derivatives, and oligomers. X halogen atom or hydrogen, R methyl or hydrogen

diols (Fig. 2.4) (Hudlicky et al. 1993). No further breakdown has been described and therefore the chlorostyrenes serve not as carbon source for strain 39D. Another possible biotransformation has been revealed by several other studies which demonstrated that SMOs are able to epoxidize halogenated styrenes (reviewed by Montersino et al. 2011; see Sects. 4.1 and 5.1). Additionally, the p-cymene monooxygenase obtained of *P. putida* F1 converts 4-chlorostyrene into the 4-chlorostyrene oxide (Nishio et al. 2001). More recently, it has been shown to take advantage of the whole side-chain oxidation pathway in order to convert halogenated styrenes into corresponding phenylacetic acids (Oelschlägel et al. 2014a). Several microorganisms (e.g., *R. opacus* 1CP and *P. fluorescens* ST) can be used to convert styrene and substituted derivatives. Surprisingly, it has been found that even phenylacetic acid was overproduced from styrene and released into the fermentation medium. But, if halogenated styrenes were applied, no further utilization of respective halogenated phenylacetic acids was observed. Therefore, halogenated styrene derivatives were not biodegraded, but metabolites of bioconversions were enriched (Fig. 2.4).

Styrene oligomers can be degraded by soil microorganisms to some extent as reported by Tuschii and coworkers (1977). The dimer disappears rapidly from mixed cultures and a degradation of up to 70 % was determined. But, already the trimer appeared to be a poor substrate since less than 15 % could be degraded, whereas higher oligomers were not transformed by the soil microorganisms. During this study, the strain *Alcaligenes* sp. 559 was isolated from the consortium mentioned and found to utilize such styrene oligomers as sole carbon source. Further, the strain 559 was able to use biphenyl and *trans*-stilbene, but not the monomer styrene. The latter is a clear indication that the microbial degradation of styrene oligomers does not involve enzymes of the styrene-catabolic pathways described above. Later, another strain described as *Pseudomonas* sp. 419 was also found to degrade styrene dimers (Higashimura et al. 1983).

2.5 Styrene Transformation in Humans

The monomer styrene is mainly taken up via lungs, skin, or rarely by an oral application (Hartmans 1995; Khaksar and Ghazi-Khansaris 2009; Rueff et al. 2009; Warhurst and Fewson 1994). It is rapidly distributed among the organs and usually attacked at the vinyl side-chain to yield styrene oxide prior further transformation (Rueff et al. 2009). The respective enzyme involved in epoxidation is a cytochrome P450 monooxygenase (CYP). And it is supposed that the reactive epoxide formed is mainly causing the (geno)toxic effects related to styrene exposure. This metabolite can be hydrolyzed by EH yielding styreneglycol which can be further oxidized by means of dehydrogenases. Thus mandelic acid and phenylglyoxylic acid are produced and represent the main urinary metabolites after styrene transformation takes place in human body. Also the transamination of these metabolites is possible which finally yields corresponding amino acids and phenylglycine. A similar biotransformation of styrene is carried out by some fungi (Fig. 2.1) (Braun-Lüllemann et al. 1997).

References

Alexandrino M, Knief C, Lipski A (2001) Stable-isotope-based labeling of styrene-degrading microorganisms in biofilters. Appl Environ Microbiol 67:4796–4804

Araya A, Chamy R, Mota M, Alves M (2000) Biodegradability and toxicity of styrene in the anaerobic digestion process. Biotechnol Lett 22:1477–1481

Arnold M, Reittu A, Von Wright A, Martikainen PJ, Suihko ML (1997) Bacterial degradation of styrene in waste gases using a peat filter. Appl Microbiol Biotechnol 48:738–744

Baggi G, Boga MM, Catelani D, Galli E, Treccani V (1983) Styrene catabolism by a strain of *Pseudomonas fluorescens*. Syst Appl Microbiol 4:141–147

Beltrametti F, Marconi AM, Bestetti G, Galli E, Ruzzi M, Zennaro E (1997) Sequencing and functional analysis of styrene catabolism genes from *Pseudomonas fluorescens* ST. Appl Environ Microbiol 63:2232–2239

Bestetti G, Galli E, Ruzzi M, Baldacci G, Zennaro E, Frontali L (1984) Molecular characterization of a plasmid from *Pseudomonas fluorescens* involved in styrene degradation. Plasmid 12:181–188

Bestetti G, Galli E, Benigni C, Orsini F, Pelizzoni F (1989) Biotransformation of styrenes by a *Pseudomonas putida*. Appl Microbiol Biotechnol 30:252–256

Braun-Lüllemann A, Majcherczyk A, Huttermann A (1997) Degradation of styrene by white-rot fungi. Appl Microbiol Biotechnol 47:150–155

Burback BL, Perry JJ (1993) Biodegradation and biotransformation of groundwater pollutant mixtures by *Mycobacterium vaccae*. Appl Environ Microbiol 59:1025–1029

Carmona M, Zamarro MT, Blázquez B, Durante-Rodríguez G, Juárez JF, Valderrama JA, Barragán MJL, García JL, Diaz E (2009) Anaerobic catabolism of aromatic compounds: a genetic and genomic view. Microbiol Mol Biol Rev 73:71–133

Ceccoli RD, Bianchi DA, Rial DV (2014) Flavoprotein monooxygenases for oxidative biocatalysis: recombinant expression in microbial hosts and applications. Frontiers Microbiology 5:1–14

Chen X, Kohl TA, Rückert C, Rodionov DA, Li L-H, Ding J-Y, Kalinowski J, Liu S-J (2012) Phenylacetic acid catabolism and its transcriptional regulation in *Corynebacterium glutamicum*. Appl Environ Microbiol 78:5796–5804

Cho MC, Kang D-O, Yoon BD, Lee K (2000) Toluene degradation pathway from *Pseudomonas putida* F1: substrate specificity and gene induction by 1-substituted benzenes. J Ind Microbio Biotech 25:163–170

Colby J, Stirling DI, Dalton H (1977) The soluble methane monooxygenase of *Methylococcus capsullatus* (Bath). Biochem J 165:395–402

Cox HH, Faber BW, Heiningen WNV, Radhoe H, Doddema HJ, Harder W (1996) Styrene metabolism in *Exophiala jeanselmei* and involvement of a cytochrome P-450-dependent styrene monooxygenase. Appl Environ Microbiol 62:1471–1474

Cox HH, Moerman RE, van Baalen S, van Heiningen WN, Doddema HJ, Harder W (1997) Performance of a styrene-degrading biofilter containing the yeast *Exophiala jeanselmei*. Biotechnol Bioeng 53:259–266

Cox HHJ (1995) Styrene removal from waste gas by the fungus *Exophiala jeanselmei* in a biofilter. Ph.D. thesis, University of Groningen, The Netherlands

Cox HHJ, Houtman JHM, Doddema HJ, Harder W (1993) Enrichment of fungi and degradation of styrene in biofilters. Biotechnol Lett 15:737–742

de Jong E, Beuling EE, van der Zwan RP, de Bont JAM (1990) Degradation of veratryl alcohol by *Penicillium simplicissimum*. Appl Microbiol Biotechnol 34:420–425

del Peso-Santos T, Bartolomé-Martín D, Fernández C, Alonso S, García JL, Díaz E, Shingler V, Perera J (2006) Coregulation by phenylacetyl-Coenzyme A-responsive PaaX integrates control of the upper and lower pathways for catabolism of styrene by *Pseudomonas* sp. strain Y2. J Bacteriol 188:4812–4821

Furuhashi K, Shintani M, Takagi M (1986) Effects of solvents on the production of epoxides by Nocardia coralline B-276. Appl Microbiol Biotechnol 32:218–223

Gąszczak A, Bartelmus G, Greń I (2012) Kinetics of styrene biodegradation by *Pseudomonas* sp. E-93486. Appl Microbiol Biotechnol 93:565–573

Geigert J, Lee TD, Dalietos DJ, Hirano DS, Neidleman SL (1986) Epoxidation of alkenes by chloroperoxidase catalysis. Biochem Biophys Res Comm 136:778–782

Grbić-Galić D, Churchman-Eisel N, Mraković I (1990) Microbial transformation of styrene by anaerobic consortia. J Appl Bacteriol 69:247–260

Han JH, Park MS, Bae JW, Lee EY, Yoon YJ, Lee S-G, Park S (2006) Production of (S)-styrene oxide using styrene oxide isomerase negative mutant of *Pseudomonas putida* SN1. Enzyme Microb Technol 39:1264–1269

Hartmans S (1995) Microbial degradation of styrene. In: Biotransformations: microbiological degradation of health risk compounds. Elsevier Science 32:227–238

Hartmans S, Smits JP, van der Werf MJ, Volkering F, de Bont JAM (1989) Metabolism of styrene oxide and 2-phenylethanol in the styrene-degrading *Xanthobacter* strain 124X. Appl Environ Microbiol 55:2850–2855

Hartmans S, van der Werf MJ, De Bont JAM (1990) Bacterial degradation of styrene involving a novel flavin adenine dinucleotide-dependent styrene monooxygenase. Appl Environ Microbiol 56:1347–1351

Higashimura T, Sawamoto M, Hiza T, Karaiwa M, Tsuchii A, Suzuki T (1983) Effect of methyl substitution on microbial degradation of linear styrene dimers by two soil bacteria. Appl Environ Microbiol 46:386–391

Higgins IJ, Hammond RC, Sariaslani FS, Best D, Davies MM, Tryhorn SE, Taylor F (1979) Biotransformation of hydrocarbons and related compounds by whole organism suspensions of methane-grown *Methylosinus trichosporium* OB3b. Biochem Biophys Res Comm 89:671–677

Hou CT, Patel R, Laskin AI, Barnabe N, Barist I (1983) Epoxidation of short-chain alkenes by resting-cell suspensions of propane-grown bacteria. Appl Environ Microbial 46:171–177

Hudlicky T, Boros EE, Boros CH (1993) New diol metabolites derived by bioxidation of chlorostyrenes with *Pseudomonas putida*: determination of absolute stereochemistry and enantiomeric excess by convergent syntheses. Tetrahedron Asymmetry 4:1365–1386

Huijbers MME, Montersino S, Westphal AH, Tischler D, van Berkel WJH (2014) Flavin dependent monooxygenases. Arch Biochem Biophys 544:2–17

Ikura Y, Yoshida Y, Kudo T (1997) Physiological properties of two *Pseudomonas mendocina* strains which assimilate styrene in a two-phase (solvent-aqueous) system under static culture conditions. J Ferment Bioeng 83:604–607

Itoh N, Hayashi K, Okada K, Ito T, Mizuguchi N (1997) Characterization of styrene oxide isomerase, a key enzyme of styrene and styrene oxide metabolism in *Corynebacterium* sp. Biosci Biotech Biochem 61:2058–2062

Itoh N, Yoshida K, Okada K (1996) Isolation and identification of styrene-degrading *Corynebacterium* strains, and their styrene metabolism. Biosci Biotechnol Biochem 60:1826–1830

Jung I-G, Park C-H (2005) Characteristics of styrene degradation by *Rhodococcus pyridinovorans* isolated from a biofilter. Chemosphere 61:451–456

Kantz A, Chin F, Nallamothu N, Nguyen T, Gassner GT (2005) Mechanism of flavin transfer and oxygen activation by the two-component flavoenzyme styrene monooxygenase. Arch Biochem Biophys 442:102–116

Keener WK, Arp DJ (1994) Transformation of aromatic compounds by *Nitrosomonas europaea*. Appl Environ Microbiol 60:1914–1920

Khaksar M-R, Ghazi-Khansari M (2009) Determination of migration monomer styrene from GPPS (general purpose polystyrene) and HIPS (high impact polystyrene) cups to hot drinks. Toxicol Mech Method 19:257–261

Kim J, Ryu HW, Jung DJ, Lee TH, Cho K-S (2005) Styrene degradation in a polyurethane biofilter inoculated with *Pseudomonas* sp. IS-3. J Microbiol Biotechnol 15:1207–1213

Knackmuss H-J, Hellwig M, Lackner H, Otting W (1976) Cometabolism of 3-methylbenzoate and methylcatechols by a 3-chlorobenzoate utilizing *Pseudomonas*: accumulation of (+)-2,5-dihydro-4-methyl- and (+)-2,5- dihydro-2-methyl-5-oxo-furan-2-acetic acid. Eur J Appl Microbiol 2:267–276

Lin H, Qiao J, Liu Y, Wu Z-L (2010) Styrene monooxygenase from *Pseudomonas* sp. LQ26 catalyzes the asymmetric epoxidation of both conjugated and unconjugated alkenes. J Mol Catal B Enzym 67:236–241

Lu C, Lin M-R, Lin J (2001) Removal of styrene vapor from waste gases by a trickle-bed air biofilter. J Hazard Mater B82:233–245

Marconi AM, Beltrametti F, Bestetti G, Solinas F, Ruzzi M, Galli E, Zennaro E (1996) Cloning and characterization of styrene catabolism genes from *Pseudomonas fluorescens* ST. Appl Environ Microbiol 62:121–127

Marín M, Pérez-Pantoja D, Donoso R, Wray V, González B, Pieper DH (2010) Modified 3-oxoadipate pathway for the biodegradation of methylaromatics in *Pseudomonas reinekei* MT1. J Bacteriol 192:1543–1552

Mars AE, Kasberg T, Kaschabek SR, van Agteren MH, Janssen DB, Reineke W (1997) Microbial degradation of chloroaromatics: use of the *meta*-cleavage pathway for mineralization of chlorobenzene. J Bacteriol 179:4530–4537

Miyamoto K, Okuro K, Ohta H (2007) Substrate specificity and reaction mechanism of recombinant styrene oxide isomerase from *Pseudomonas putida* S12. Tetrahedron Lett 48:3255–3257

Montersino S, Tischler D, Gassner GT, van Berkel WJH (2011) Catalytic and structural features of flavoprotein hydroxylases and epoxidases. Adv Synth Catal 353:2301–2319

Mooney A, O'Leary ND, Dobson ADW (2006a) Cloning and functional characterization of the *styE* gene, involved in styrene transport in *Pseudomonas putida* CA-3. Appl Environ Microbiol 72:1302–1309

Mooney A, Ward PG, O'Connor KE (2006b) Microbial degradation of styrene: biochemistry, molecular genetics, and perspectives for biotechnological applications. Appl Microbiol Biotechnol 72:1–10

Navarro-Llorens JM, Patrauchan MA, Stewart GR, Davies JE, Eltis LD, Mohn WW (2005) Phenylacetate catabolism in *Rhodococcus* sp. strain RHA1: a central pathway for degradation of aromatic compounds. J Bacteriol 187:4497–4504

Nikodinovic-Runic J, Flanagan M, Hume AR, Cagney G, O'Connor KE (2009) Analysis of the *Pseudomonas putida* CA-3 proteome during growth on styrene under nitrogen-limiting and non-limiting conditions. Microbiology 155:3348–3361

Nishio T, Patel A, Wang Y, Lau PC (2001) Biotransformations catalyzed by cloned *p*-cymene monooxygenase from *Pseudomonas putida* F1. Appl Microbiol Biotechnol 55:321–325

O'Connor K, Buckley CM, Hartmans S, Dobson AD (1995) Possible regulatory role for non-aromatic carbon sources in styrene degradation by *Pseudomonas putida* CA-3. Appl Environ Microbiol 61:544–548

O'Connor KE, Dobson AD, Hartmans S (1997) Indigo formation by microorganisms expressing styrene monooxygenase activity. Appl Environ Microbiol 63:4287–4291

O'Leary ND, O'Connor KE, Dobson ADW (2002) Biochemistry, genetics and physiology of microbial styrene degradation. FEMS Microbiol Rev 26:403–417

Oelschlägel M, Gröning JAD, Tischler D, Kaschabek SR, Schlömann M (2012) Styrene oxide isomerase of *Rhodococcus opacus* 1CP, a highly stable and considerably active enzyme. Appl Environ Microbiol 78:4330–4337

Oelschlägel M, Kaschabek SR, Zimmerling J, Schlömann M, Tischler D (2015) Co-metabolic formation of substituted phenylacetic acids by styrene-degrading bacteria. Biotechnol Rep (Amst.) 6:20–26

Oelschlägel M, Zimmerling J, Tischler D, Schlömann M (2014a) Method for biocatalytic synthesis of substituted or unsubstituted phenylacetic acids and ketones having enzymes of microbial styrene degradation. Patent: DE 102013211075 A1 20141218; WO 2014198871 A2 20141218

Oelschlägel M, Zimmerling J, Schlömann M, Tischler D (2014b) Styrene oxide isomerase of *Sphingopyxis* sp. Kp5.2. Microbiology 160:2481–2491

Omori T, Jigami Y, Minoda Y (1974) Microbial oxidation of α-methylstyrene and β-methylstyrene. Agr Biol Chem 38:409–415

Olivera ER, Miñambres B, García B, Muñiz C, Moreno MA, Ferrández A, Díaz E, García JL, Luengo JM (1998) Molecular characterization of the phenylacetic acid catabolic pathway in *Pseudomonas putida* U: the phenylacetyl-CoA catabolon. Proc Natl Acad Sci USA 95:6419–6424

Paca J, Koutsky B, Maryska M, Halecky M (2001) Styrene degradation along the bed height of perlite biofilter. J Chem Technol Biotechnol 76:873–878

Panke S, Witholt B, Schmid A, Wubbolts MG (1998) Towards a biocatalyst for (*S*)-styrene oxide production: characterization of the styrene degradation pathway of *Pseudomonas* sp. strain VLB120. Appl Environ Microbiol 64:2032–2043

Park MS, Bae JW, Han JH, Lee EY, Lee S-G, Park S (2006) Characterization of styrene catabolic genes of *Pseudomonas putida* SN1 and construction of a recombinant *Escherichia coli* containing styrene monooxygenase gene for the production of (*S*)-styrene oxide. J Microbiol Biotechnol 16:1032–1040

Patrauchan MA, Florizone C, Eapen S, Gómez-Gil L, Sethuraman B, Fukuda M, Davies J, Mohn WW, Eltis LD (2008) Roles of ring-hydroxylating dioxygenases in styrene and benzene catabolism in *Rhodococcus jostii* RHA1. J Bacteriol 190:37–47

Przybulewska K, Wieczorek A, Nowak A (2006) Isolation of microorganisms capable of styrene degradation. Polish J Environ Stud 15:777–783

René ER, Veiga MC, Kennes C (2010) Biodegradation of gas-phase styrene using the fungus *Sporothrix variecibatus*: impact of pollutant load and transient operation. Chemosphere 79:221–227

Rueff J, Teixeira JP, Santos LS, Gaspar JF (2009) Genetic effects and biotoxicity monitoring of occupational styrene exposure. Clin Chim Acta 399:8–23

Rustemov SA, Golovleva LA, Alieva RM, Baskunov BP (1992) New pathway of styrene oxidation by a *Pseudomonas putida* culture. Microbiologica 61:1–5

Shirai K, Hisatsuka K (1979) Production of β-phenethyl alcohol from styrene by *Pseudomonas* 305-STR-1-4. Agric Biol Chem 43:1399–1406

Smith MR (1990) The biodegradation of aromatic hydrocarbons by bacteria. Biodegradation 1:191–206

Teufel R, Mascaraque V, Ismail W, Voss M, Perera J, Eisenreich W, Haehnel W, Fuchs G (2010) Bacterial phenylalanine and phenylacetate catabolic pathway revealed. Proc Natl Acad Sci USA 107:14390–14395

Tischler D, Eulberg D, Lakner S, Kaschabek SR, van Berkel WJH, Schlömann M (2009) Identification of a novel self-sufficient styrene monooxygenase from *Rhodococcus opacus* 1CP. J Bacteriol 191:4996–5009

Tischler D, Kaschabek SR (2012) Microbial degradation of xenobiotics. In: Singh SN (ed). Springer, Berlin, pp 67–99

Toda H, Itoh N (2012) isolation and characterization of styrene metabolism genes from styrene-assimilating soil bacteria *Rhodococcus* sp. ST-5 and ST-10. J Biosci Bioeng 113:12–19

Tuschii A, Suzuki T, Takahara Y (1977) Microbial degradation of styrene oligomer. Agric Biol Chem 41:2417–2421

Utkin I, Yakimov M, Matveeva L, Kozlyak E, Rogozhin I, Solomon Z, Bezborodov A (1991) Degradation of styrene and ethylbenzene by *Pseudomonas* species Y2. FEMS Microbiol Lett 77:237–242

Velasco A, Alonso S, Garcia JL, Perera J, Diaz E (1998) Genetic and functional analysis of the styrene catabolic cluster of *Pseudomonas* sp. strain Y2. J Bacteriol 180:1063–1071

Warhurst AM, Clarke KF, Hill RA, Holt RA, Fewson CA (1994a) Metabolism of styrene by *Rhodococcus rhodochrous* NCIMB 13259. Appl Environ Microbiol 60:1137–1145

Warhurst AM, Clarke KF, Hill RA, Holt RA, Fewson CA (1994b) Production of catechols and mucinic acids from various aromatics by the styrene-degrader *Rhodococcus rhodochrous* NCIMB 13259. Biotechnol Lett 16:513–516

Warhurst AM, Fewson CA (1994) A review. Microbial metabolism and biotransformation of styrene. J Appl Bacteriol 77:597–606

Weber FJ (1995) Toluene: biological waste gas treatment, toxicity and microbial adaption. PhD thesis, Wageningen University, The Netherlands

Weber FJ, Hage KC, de Bont JAM (1995) Growth of he fungus Cladiosporium sphaerospermum with toluene as sole carbon and energy source. Appl Environ Microbiol 61:3562–3566

Weigner P, Páca J, Loskot P, Koutský B, Sobotka M (2001) The start-up period of styrene degrading biofilters. Folia Microbiol 46:211–216

Zilli M, Converti A, Di Felice R (2003) Macrokinetic and quantitative microbial investigation on a bench-scale biofilter treating styrene-polluted gaseous streams. Biotechnol Bioeng 83:29–38

Chapter 3
Molecular Genetics of Styrene Degrading Routes

Abstract Various pathways for the microbial catabolism of styrene were described where only the side-chain oxidation route seems to be specific for styrene. Therefore, we have to discriminate between those different routes and herein the focus is mainly on the styrene-specific one. Styrene is activated and oxidized to phenylacetic acid by the proteins encoded within a so-called *sty(rene)*-operon. This operon comprises the genes coding for styrene monooxygenase (*styA/styB*: SMO), styrene oxide isomerase (*styC*: SOI), and phenylacetaldehyde dehydrogenase (*styD*: PAD). But, the molecular genetic background is limited mainly to pseudomonads. Next to the genes coding for styrene metabolizing enzymes, transporter (*styE*) and regulatory elements (*styS/styR*) are also encoded. In few cases, a proof of function was shown and will be discussed. Further, effect of phenylacetic acid pathway and its regulatory machinery were found to regulate (repress) the *sty*-operon. However, there were also some contradictions determined as for example, side-chain activity but incomplete *sty*-gene cluster. It indicates different genetic organization or even alternative pathway-enzymes.

Keywords Styrene operon · Upper and lower styrene pathway · Two-component sensor kinase · Transcriptional repressor · *styABCD* · *styE* · *stySR*

3.1 The *Sty*-Operon of the Upper Styrene Degradation Pathway

Earlier studies aimed to identify the styrene catabolic active enzymes and describe the respective pathways and metabolites (Fig. 3.1a). Thus different *Pseudomonas* species (Beltrametti et al. 1997; Kantz et al. 2005; Lin et al. 2010; Marconi et al. 1996, O'Connor et al. 1995, 1997; O'Leary et al. 2001, 2002a; Otto et al. 2004;

© The Author(s) 2015
D. Tischler, *Microbial Styrene Degradation*,
SpringerBriefs in Microbiology, DOI 10.1007/978-3-319-24862-2_3

23

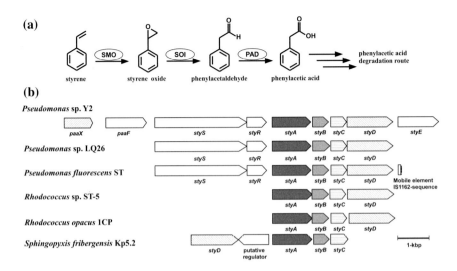

Fig. 3.1 The upper styrene degradation pathway converts styrene to phenylacetic acid (**a**) which is reflected by a conserved genetic background (**b**). The *sty*-operon comprises regulatory elements (*styS/styR*), catabolic active enzymes (*styABCD*), and encodes for a transporter (*styE*). In view cases, this operon is located next to the lower styrene pathway which is in fact the general phenylacetic acid route (*paa*-genes). Here, the initial activity of a phenylacetyl-CoA ligase activates the central intermediate for further degradation. Only a limited selection of representative operons are shown here. (It was adapted according to Tischler and Kaschabek 2012)

Panke et al. 1998; Utkin et al. 1991; Velasco et al. 1998) and some *Rhodococcus* strains (Oelschlägel et al. 2012; Tischler et al. 2009; Toda and Itoh 2012) were extensively studied for their ability to mineralize styrene. Very recently, the upper styrene pathway of *Sphingopyxis fribergensis* Kp5.2 was investigated and found to be differently organized then the others mentioned (Oelschlägel et al. 2014, 2015). Later, those strains were investigated also from a molecular genetic attitude. Respectively, most of them harbor a complete and functional *sty*-operon (Fig. 3.1b). Indications for a coordinated expression of *styABCD* as a single mRNA were reported (O'Leary et al. 2001; Santos et al. 2000; Velasco et al. 1998). The latter means a polycistronic mRNA is produced from the *sty*-operon and processed to yield finally the catabolic active enzymes.

The *sty*-operon comprises four catabolic active enzymes. The first two genes (*styA/styB*) code for the two-component styrene monooxygenase (SMO) (Huijbers et al. 2014). Subunit StyB utilizes NADH to reduce FAD which is transferred to the subunit StyA, and after activation of molecular oxygen in the form of a C4α-(hydro-)peroxy-FAD styrene can be epoxized. The epoxide formed is a substrate to the SOI which is encoded by *styC* (Itoh et al. 1997). This biocatalyst represents an intramolecular oxidoreductase and produces phenylacetaldehyde which subsequently needs to be detoxified which is achieved by a phenylacetaldehyde dehydrogenase (PAD) (Velasco et al. 1998). That is the last catabolic enzyme of the operon and encoded by *styD*. Furthermore, in the most strains a two-component regulatory system is present and encoded upstream to *styABCD*. The genes *styS* and *styR*, respectively, code for StyS (sensor kinase) and StyR (response regulator) which is a positive regulator of the *sty*-operon, inducing transcription upon styrene presence (see Sect. 3.2). However, this regulatory element has not been determined on chromosomal level in case of *Rhodococcus* sp. ST-5 (Toda and Itoh 2012). In case of *Sphingopyxis fribergensis* Kp5.2 also, a different regulatory element has been proposed (Oelschlägel et al. 2014; Fig. 3.3); but, proof of activity need to be demonstrated. But, a strictly regulated *sty*-gene expression pattern of strain ST-5 was observed and indicates a similar positive regulation as in pseudomonads. The latter still needs to be shown for rhodococci and respective regulatory genes have to be elucidated.

In addition to the mentioned catabolic and regulatory elements in a few pseudomonads, an ATPase-like transporter designated as StyE is encoded by *styE* downstream to *styD* (Mooney et al. 2006; Velasco et al. 1998). On protein level, it shows significant similarity to other membrane-linked transport proteins and therewith a putative role as an active styrene transporter was suggested. Interestingly, a sequence similarity of about 50 % on amino acid level to structurally investigated transport porins of toluene is predicted from comparative BLAST analyses. Further, a styrene-dependent transcription of *styE* and *sty*-operon coexpression reinforces this hypothesis (Mooney et al. 2006). By increasing the copies of *styE* and therewith its expression level in *Pseudomonas putida* CA-3, enhanced styrene monooxygenase transcription level (8-fold) and activity (4-fold) were observed. A general ATPase inhibitor as vanadate revealed to be less effective when StyE was overproduced. All together, these observations indicate an active role of StyE in styrene import by *Pseudomonas* cells. However, it needs to be mentioned that contrary results were later obtained for the same strain (Nikodinovic-Runic et al. 2009). A proteomic approach revealed only minor StyE amounts from differently cultivated strain CA-3. From these observations, various conclusions can be drawn. First, StyE might be a highly efficient transporter and thus low levels are sufficient enough. Second, the cell likes to prevent extensive uptake of styrene and probable toxic side reactions. Third, other transport proteins can support or even facilitate the styrene import as well. And fourth, of course the transmembrane diffusion of styrene needs to be considered. Especially, the diffusion of styrene into cells is likely (see Chap. 1) and the low abundance of *styE*-like genes among reported styrene-degraders confirms the necessity of additional entrance routes for styrene.

3.2   Regulatory Elements of the *Sty*-Operon and *Sty*-Gene Expression

A number of cellular processes are regulated by two-component signal transduction systems; for example, metabolic and transport activities of prokaryotes (Bijlsma and Groisman 2003; Milani et al. 2005; Reizer and Saier 1997).

These regulators are generally composed of a sensor histidine kinase (HK) and a response regulator (RR). According to a proper induction (e.g., chemical signal), the HK gets autophosphorylated at a conserved histidine residue. The reaction consumes ATP and releases ADP which represents a kinase activity. Subsequently, the high-energy phosphoryl group is transferred to a conserved aspartate at the N-terminal domain of the RR. This phosphorylation activates the C-terminal domain of the RR component which is often a transcription factor. Due to this transfer reaction, the DNA-binding properties of the RR get changed and thus regulate transcription of adjacent genes. Such two-component regulatory elements have been described, for example, in toluene (Coshigano and Young 1997; Lau et al. 1997; Leuthner and Heider 1998) and biphenyl degradation (Labbé et al. 1997), but, also in styrene catabolism of *Pseudomonas* strains (Velasco et al. 1998). Interestingly, a close relation of respective regulators was demonstrated since styrene acts as an inducer (chemical signal) for the HK of toluene degradation (Cho et al. 2000; Mosqueda and Ramos 2000). This is another connection to toluene degradation as it was described above for the transporter porin StyE which also has a related counterpart in toluene degradation.

The *sty*-operon of pseudomonads comprises such a two-component regulator which is encoded by *styS* and *styR* (Fig. 3.2a) (Leoni et al. 2003; Massai et al. 2014; Milani et al. 2005; O'Leary et al. 2001, 2002a, b; Santos et al. 2000; Velasco et al. 1998). The respective gene products (StyS = HK, and StyR = RR) display significant similarities to a number of two-component signal transduction systems (Reizer and Saier 1997). Usually, these two genes are found to be located upstream of the *styABCD*(*E*)-genes and are supposed to be transcribed together (O'Leary et al. 2001). Thus, they regulate the gene expression of proteins involved in styrene uptake (StyE) as well as catabolism (StyABCD).

3.2.1 Proteins of the Two-Component Regulator StySR

The two components of such regulatory elements are modular and the different domains are responsible for different parts of the regulation process (Bijlsma and Groisman 2003; Reizer and Saier 1997). The sensor components (HK) can vary in size and composition. A single sensing domain can be connected to a single kinase domain and pursue the transfer of high-energy phosphoryl group onto the RR initiating the regulation along with gene transcription. However, additional phosphoryl group transferring proteins can be involved. Thus, separate proteins or multidomain hybrid sensors occur in nature.

StyS (about 110 kDa) represents a multidomain sensor protein comprising five domains as follows: input-1, histidine kinase-1 (HK-1), receiver, input-2, and histidine kinase-2 (HK-2). Similar HKs can be found among the regulatory elements of toluene degradation pathway (TodS, TutC, and TutS) (Coshigano and Young 1997; Lau et al. 1997; Leuthner and Heider 1998). The input domains of StyS contain conserved motives of PAS-sensing domains which are general sensor

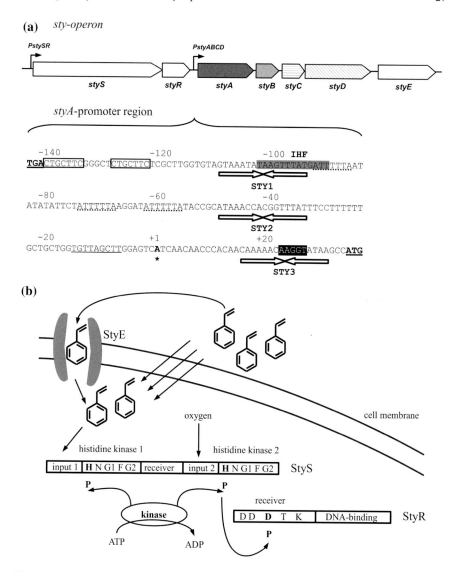

Fig. 3.2 The regulatory machinery of the *sty*-operon is (**a**) shown on genetic level (compare Fig. 3.1a, b) as a comic to highlight the sensing path from styrene uptake toward activation of the DNA-binding protein StyR (modified according to Tischler and Kaschabek 2012). The two *sty*-operon-inclusive promoters are highlighted while in detail the promotor region next to *styA* is outlined (O'Leary et al. 2002b). The numbering is given according to the transcription start (*asterisk*). Upon styrene sensing, the C-terminal domain of StyR can bind to STY-sites, especially STY2, allowing RNA polymerase binding to the underlined promotor region which initiates the transcription process and thus the production of StyABCD. A detailed description of all elements is given in the text (see Sects. 3.2.1 and 3.2.2)

modules for sensing the changes of redox states, light and small ligand concentrations (e.g. styrene or its metabolites). Therefore, it is supposed that styrene is detected by one or both input domains as a result of altered redox potential or as a ligand binding to the protein (Milani et al. 2005; Santos et al. 2000; Taylor and Zhulin 1999; Velasco et al. 1998). But, the exact mechanism of styrene sensing still needs to be uncovered. However, after the styrene determination by the sensor, the histidine kinase activity comes into play. The two histidine kinase domains are highly conserved and harbor the essential amino acid motif H-N-G-F-G (Parkinson and Kofoid 1992), whereas the histidine residue is necessary for the phosphotransfer event toward StyR (Bijlsma and Groisman 2003; Reizer and Saier 1997). These domains can be classified into kinase superfamilies 1a (HK-1) and 4 (HK-2), respectively. Interestingly, two modules of input domains and histidine kinases with styrene sensor and autokinase activity, respectively, are separated by an internal receiver domain. This belongs to the RA2-receiver protein superfamily and includes a conserved motif (D-D-S-K) (Grebe and Stock 1999). Finally, StyS can sense the amount of styrene as well as the redox environment inside the cell and convert it into a signal of a high-energy phosphoryl group. This mechanism controls the level of StyR-phosphorylation according to the environmental or cytoplasm conditions (Milani et al. 2005).

StyR (about 24 kDa) is the response subunit of the *sty*-regulatory system and has counterparts in further aromatic degradation pathways as TodT and TudC (Lau et al. 1997; Leuthner and Heider 1998). It can be classified into the protein superfamily of FixL/NarL RRs and comprises respective domains. At the N-terminal side, a receiver domain is located (amino acid position 1–127). This receiver belongs to the RA4-subfamily and can be recognized by a highly conserved amino acid motif (D-D-D-T-K) (Baikalov et al. 1996; Grebe and Stock 1999). The central aspartate represents the phosphoryl group acceptor. At the C-terminal side, a regulatory domain is located (amino acid position 142–208). This is the DNA-binding domain and has similarities to other family-3 RR as NarL. Interestingly, a 34-amino acid-long Q-linker connects both mentioned domains (O'Leary et al. 2002b) which has an α-helical structure and is a rare feature among related RRs (Milani et al. 2005). Furthermore, structural studies imply that StyR occurs in an equilibrium state and upon phosphorylation it forms a dimer which can bind to DNA (Milani et al. 2005). Finally, the amount of StyR present in the cell and its degree of phosphorylation controls the *styABCD(E)*-gene expression (Leoni et al. 2005) and therefore the catabolic activity.

An overview about different domains of the two-component regulator StySR and their function is displayed in Fig. 3.2.

3.2.2 Styrene-Induced Transcription of Sty-Genes

Expression of styrene catabolic genes (*styABCD*) is dependent on the presence of styrene (Nikodinovic-Runic et al. 2009; Velasco et al. 1998) and often transcripts of respective regulatory genes (*styS* and *styR*) were determined as well (O'Connor et al. 1995; O'Leary et al. 2001). Thus, a major role of StySR in *sty*-operon regulation was demonstrated and will be discussed herein (Leoni et al. 2003, 2005; O'Leary et al. 2001, 2002a; Panke et al. 1998; Santos et al. 2000; Velasco et al. 1998). However, repression of *sty*-genes was reported, especially, when other potential carbon sources had been supplied. For example, glucose, glutamate, citrate, and even more striking the product of StyABCD-activity phenylacetic acid can repress the transcription of *sty*-genes (O'Connor et al. 1995; O'Leary et al. 2001; Santos et al. 2000). Even the addition of styrene does not automatically prevent this repression. Since styrene is the major signal molecule for initialization of the *sty*-gene transcription and not phenylacetic acid or its metabolites, it can be reasoned that the upper styrene pathway is independently regulated from the lower styrene pathway (phenylacetyl-CoA catabolon; see Sects. 2.2.2 and 3.5.2). However, later studies provided evidence for a more general control mechanism (Peso-Santos et al. 2006). The PaaX-repressor of the phenylacetyl-CoA catabolon regulates phenylacetic acid catabolism and in addition it does bind to a sequence in the *styA*-gene and therewith repress its expression. This mechanism indicates a PaaX-controlled transcription of *sty*-genes as an additional regulatory element for both upper and lower styrene degradation pathway.

In order to uncover and understand the regulatory elements of the *sty*-operon, several studies in *Escherichia coli* with fragments obtained of *Pseudomonas* species Y2 and ST were performed. The following mechanism represents the key steps toward *sty*-gene transcription or repression (Fig. 3.2). The expression of *styS* and *styR* was found to be constitutive in *Pseudomonas fluorescens* ST and independent of the carbon source supplied (Santos et al. 2000). This observation indicates the respective promoter *PstySR* is not subject to catabolite repression and in general low levels of the regulatory elements StySR are present. However, the promoter *PstyABCD* of the *sty*-genes was found to be induced by styrene and repressed to various extents by other carbon sources supplied.

In the case of solely suppling styrene as carbon and energy source, an inductive and active regulation has been determined. Styrene is sensed by one of the input domains of StyS leading to an autokinase activity consuming ATP and a histidine of one of the histidine kinase domains of StyS gets simultaneously phosphorylated. This activation allows subsequently transferring the phosphoryl group onto the conserved aspartate residue of StyR receiver domain and induces dimer formation (Leoni et al. 2003; Milani et al. 2005; Santos et al. 2002). Thereafter, the DNA-binding is enhanced and likely to take place at the STY2 sequence. The latter is a palindromic sequence part of the *styA* promoter region (*PstyABCD*; Fig. 3.2). Further, it is highly similar to the respective region of a toluene operon to which the corresponding RR (TodT) binds (Lau et al. 1997). Upon this event, the RNA polymerase binding is highly attracted toward the -10 sequence of the promoter region (5'-TGTTAGCTT-3') and initiates the transcription of *sty*-genes. Therefore, StyS and StyR together act as an activator for the transcription (positive response). However, after desired gene expression increased levels of phosphorylated StyR remain in the cell and are still able to bind to the promotor region.

In that case, another binding sequence designated STY3 gets occupied as well, leading to a negative response since this event represses the transcription of *sty*-genes (Leoni et al. 2005). Further, a third potential binding sequence was identified and named STY1. The latter one seemed to have a dual function since it positively effects transcription in the presence of styrene or allows repression in case of glucose-supported growth (Leoni et al. 2005). In addition to these three promotor sequences and their StyR-interaction leading to different expression patterns, another factor was determined to effect *sty*-gene transcription in a positive manner (Santos et al. 2002). A small heterodimeric protein designated as integration host factor (IHF) can bind to a consensus sequence (5'-WATCAANNNNTTR-3'; complementary encoded to *sty*-genes) in the *styA* promoter region and effect the transcription (Leoni et al. 2005). It was suggested that upon IHF-binding, the three-dimensional structure of the promoter region changes in a way that StyR binding and transcripton are facilitated (Santos et al. 2002).

3.3 Localization and Mobility of *Sty*-Genes

First molecular genetic studies of *P. fluorescens* ST revealed the location of *sty*-genes (upper styrene pathway) on a plasmid designated pEG (Bestetti et al. 1984; Ruzzi and Zennaro 1989). Also, *Pseudomonas* sp. VLB120 was found to harbor a plasmid (pSTY) encoding for *sty*-genes (Köhler et al. 2013). Further, *Pseudomonas* sp. Y2 encodes the complete *sty*-operon on the chromosome (Velasco et al. 1998) as it was later determined for *Sphingopyxis fribergensis* Kp5.2 as well (Oelschlägel et al. 2014, 2015). Interestingly, the *sty*-operon of *P. fluorescens* ST was also found to be chromosomally encoded and not only on the pEG plasmid as reported earlier (Marconi et al. 1996). Thus, it was assumed either the complete plasmid or fragments thereof can transpose into the chromosome (Beltrametti et al. 1997; Ruzzi and Zennaro 1989). The lower styrene pathway encoded by *paa*-genes which is also known as phenylacetyl-CoA catabolon allows the phenylacetic acid metabolism which is chromosomally located (Chen et al. 2012; Luengo et al. 2001).

The circular plasmid pEG of *P. fluorescens* ST has a size of about 37 kbp and is self-transmissible (Bestetti et al. 1984). The *sty*-operon was found to be complete including both regulatory genes *styS* and *styR* (Fig. 3.1). The *styE*-gene was not identified downstream to the *sty*-genes. But, therefore sequences of transposable elements were determined (IS1162). The latter is a second time on pEG plasmid present and thus these transposable elements might explain mobility of *sty*-genes (chromosomal or plasmid) in case of strain ST. Further, the mobility of *sty*-genes among microorganisms was investigated. The strain *P. putida* PaW 340 harbors no plasmid is not able to degrade styrene. After pEG uptake by the strain PaW 340 it turned into a styrene catabolic active phenotype. This readily transfer of *sty*-genes between pseudomonads might explain the high similarity and abundance of *sty*-genes among those.

The linear megaplasmid pSTY of *Pseudomonas* sp. VLB120 (now classified as *Pseudomonas taiwanensis* VLB120; Volmer et al. 2014) has a size of about 320 kbp and codes for 365 proteins (Köhler et al. 2013). Among those are also the *sty*-genes including the regulatory elements *styS* and *styR* as well as *styABC*. However, no *styD*- or *styE*-gene could be identified and even more striking, no evidence for *sty*-genes on the chromosome was found as we reviewed earlier (Tischler and Kaschabek 2012). The absence of *styD* is in contrast to the earlier published *sty*-cluster of this strain which was obtained by conventional cloning and screening procedures (Panke et al. 1998). The different outcome needs to be clarified. Further downstream of this partial gene cluster, transposable elements are encoded on pSTY. However, from these results it can be reasoned that the *sty*-genes are present and active in strain VLB120 might have another microbial origin (mobile elements) and further the absent StyD-activity might be accomplished by other aldehyde dehydrogenases. Another interesting finding was made with respect to the phenylacetyl-CoA catabolon of strain VLB120. It harbors two *paa*-gene clusters where one is chromosomally and the other is plasmid pSTY encoded. Respectively, the *sty*-genes are next to one set of *paa*-genes as found for *Pseudomonas* sp. Y2 (Velasco et al. 1998). Also the latter strain harbors two copies of functional *paa*-gene clusters (Peso-Santos et al. 2006).

3.4 Styrene Monooxygenases Distant from *Sty*-Operons

Two types of styrene monooxygenases (SMOs) were reported according to their activity and polypeptide composition (Montersino et al. 2011; Tischler et al. 2012). However, by carefully analyzing the respective literature in combination with the genetic organization another even more complex picture develops. The genetic organization, regulation, and expression of SMO-genes of *sty*-operons were highlighted above. Therefore, solely the differences will be shown and discussed herein.

A recent study on the styrene catabolic genes of two *Rhodococcus* strains (ST-5 and ST-10) revealed two different genetic and therewith catabolic strategies (Toda and Itoh 2012). In case of strain ST-5, a functional *styABCD*-operon was identified and therefore the conventional route via side-chain oxidation was supposed and later proven by metabolite identification and enzyme assays (Fig. 3.1). In case of strain ST-10, the same metabolites were identified. But, the *styC*-gene as well as the corresponding protein for the isomerase activity on styrene oxide yielding phenylacetaldehyde had not been determined. The authors suggested a spontaneous dimerization of the epoxide to explain the still occurring phenylacetaldehyde production (Toda and Itoh 2012) which seems to be unlikely in the environment of cytoplasm. Further, no *styD*-homolog was detected in proximity of *styAB* indicating the respective gene is encoded somewhere else in the genome of strain ST-10, since the enzyme activity of StyD was determined. Respectively, in case of *Sphingopyxis fribergensis* Kp5.2, the *styD*-gene is also located at different positions in the genome in comparison to pseudomonads and their gene cluster organization (Fig. 3.1b). However, this is the first clear evidence that nature has evolved another peripheral route converting styrene into phenylacetic acid. As discussed earlier (Sect. 2.2.4), the activity of an epoxide reductase (SOR) together with an alcohol dehydrogenase (PAR) might overcome the missing SOI activity (Chap. 2, Fig. 2.1). The corresponding genes and enzyme activities need to be determined which might be straight forward since for several other microorganisms such reactions were supposed as well (Hartmans et al. 1989; Marconi et al. 1996; Shirai and Hisatsuka 1979; Utkin et al. 1991). Respectively, it means genes of such a peripheral pathway are not encoded within a gene cluster. A functional and related SMO encoded by *styA* and *styB* was also determined from metagenome (Tischler et al. 2012; van Hellemond et al. 2007). Also in this case no adjacent *styCD*-genes were determined.

Another SMO type composed of a single epoxidase (StyA1) and a naturally fused protein (StyA2B) of an epoxidase and oxidoreductase were reported (Tischler et al. 2009, 2010; van Hellemond et al. 2007). The corresponding genes seem to be part of a putative gene cluster comprising a regulatory protein (AraC like protein), α/β-hydrolase (DLH: dienelactone hydrolase or epoxide hydrolase like protein), short chain dehydrogenase/reductase (SDR), and of course the mentioned SMO-genes *styA1* and *styA2B*. A role in the styrene mineralization of strain 1CP was suggested (Tischler et al. 2009), but could not be experimentally verified so far. Interestingly, similar gene clusters are present in various microorganisms and in most cases the SMO-genes code for single epoxidase- (*styA*) and reductase- (*styB*) components (Tischler et al. 2012), whereas in most cases the functional information is lacking. In case of *Acinetobacter baylyi* ADP1, the activity of respective gene products (SMOA-/B-ADP1) was demonstrated (Gröning et al. 2014). Further, for strain ADP1 and related strains genes coding for anthranilate dioxygenase are next to those genes mentioned and might indicate a role in degradation of amino aromatic compounds (Tischler et al. 2012). However, these gene products are typical representatives acting in the aromatic compound metabolism (Braun-Lüllemann et al. 1997; Gallegos et al. 1997; Huijbers et al. 2014; Maltseva et al. 1994). Respectively, these SMO-genes code for a styrene epoxidase-like protein comprising a similar activity (Gröning et al. 2014; Tischler et al. 2009, 2010, 2013), but are most likely not involved in styrene mineralization. The statement is strengthened by an evolutionary view on the different SMO types since two phylogenetic clusters of SMOs can be classified (see Sect. 4.1) (Tischler et al. 2012).

In conclusion, two different pathways to mineralize styrene via side-chain oxidation are likely. The most prominent employs StyABCD and goes via styrene oxide, phenylacetaldehyde toward phenylacetic acid. Therefore, genetic and biochemical data are available. The second likely route replaces the StyC by SOR and PAR and therefore 2-phenylethanol appears as an additional intermediate. Here, only partial genetic and enzymatic data are available. Both of these routes employ SMOs as initial step for styrene activation, whereas the SMOs belong to the type E1 according the classification introduced by Montersino and coworkers (2011). The other SMO type (E2) is probably not involved in styrene degradation, but shows a similar enzymatic activity (Huijbers et al. 2014). Genetic organization and phylogenetic studies clearly allow differentiating those two SMO types. StyA1/StyA2B might perform also the initial activation of a styrene-like substrate via epoxidation and thus initiate the degradation of the compound.

3.5 Other Styrene Catabolic Activities and Operons

3.5.1 Direct Ring Cleavage of Styrene via Meta-Cleavage Routes

Besides the above-described *sty*-operon and its side-chain oxidation pathway also the degradation of styrene via direct ring cleavage was found (compare Chap. 2) (Warhurst et al. 1994; Patrauchan et al. 2008). Thus enzymes of biphenyl or toluene degradation might act on styrene and funnel intermediates into the metabolism. These routes allow opening the aromatic nucleus after activation by mainly *meta*-cleavage, but, also *ortho*-cleavage activity was determined (Warhurst et al. 1994). Only in few cases the latter yields no dead-end products as recently shown for a toluene utilizing *Burkholderia* strain (Dobslaw and Engesser 2014). Since various pathways might lead to a non-natural styrene activation and (partial) utilization, none of the multiple possible operons are shown here. However, it can be mentioned that here also upper genes from which encoded enzymes usually attack the aromatic ring (mono- and di-oxygenases) and lower genes from which encoded enzymes convert non-aromatic intermediates are often clustered (Arenghi et al. 2001; Kukor and Olsen 1991; Patrauchan et al. 2008) or oxygenases can be somewhat distinct located from the other clustered genes (Werlen et al. 1996). Further, as mentioned in Sect. 3.2.2 similar regulatory genes and mechanisms are involved in these aromatic degradation pathways as in the upper styrene pathway. And here the toluene pathways shows exceptional congruence.

3.5.2 The Phenylacetyl-CoA Catabolon

The central metabolite of the side-chain oxidation of styrene is phenylacetic acid which can be further metabolized by a separate pathway. Also other degradation routes yield phenylacetic acid as central metabolite, e.g., the degradation of phenylethylamine and 2-phenylethanol (Arias et al. 2008), tropic acid, ethylbenzene, or phenylacetyl amide (Luengo et al. 2001). The ability to degrade phenylacetic acid based on the *paa*-genes is not restricted to styrene-degrading bacteria. Such genes have been reported for several microorganisms, for example for *Corynebacterium glutamicum* AS1.542 (Chen et al. 2012), *E. coli* W or K-12 (Ferrández et al. 1998, Teufel et al. 2010), *Rhodococcus jostii* RHA1 (Navarro-Llorens et al. 2005), *Pseudomonas* sp. strain Y2 (Peso-Santos et al. 2006; Bartolomé-Martín et al.

2004), *P. putida* U (Olivera et al. 1998), *P. fluorescens* ST (Di Gennaro et al. 2007), and most recently for *Sphingopyxis fribergensis* Kp5.2 (Oelschlägel et al. 2015).

All steps of the phenylacetic acid degradation pathway are shown in Fig. 2.2 and described in Sects. 2.2.2. The genes encoding for the enzymes of this metabolic route are neighbored to each other and form the phenylacetic acid catabolic gene cluster. Based on the nomenclature of Teufel et al. (2010), the phenylacetic acid catabolic gene cluster comprises the genes *paaA-E* encoding ring 1,2-phenylacetyl-CoA epoxidase, the gene of the 3-oxoadipyl-CoA thiolase/3-oxo-5,6-dehydrosuberyl-CoA thiolase (*paaJ*) as well as the genes *paaG, paaZ*, and *paaK* encoding ring 1,2-epoxyphenylacetyl-CoA isomerase, oxepin-CoA hydrolase/3-oxo-5,6-dehydrosuberyl-CoA semialdehyde dehydrogenase, and phenylacetate-CoA ligase (Fig. 3.3). In some strains also a conserved thioesterase (*paaI*) is present. Furthermore, an ethyl *tert*-butyl ether degradation protein encoded by *ethD* was found in the clusters of *Rhodococcus jostii* RHA1 (Navarro-Llorens et al. 2005) and *Sphingopyxis fribergensis* Kp5.2 (Oelschlägel et al. 2015). The genes *paaF* and *paaH* encode the 2,3-dehydroadipyl-CoA hydratase and 3-hydroxyadipyl-CoA dehydrogenase. Remarkably, these genes are not present in the *paa*-gene cluster of strain Kp5.2 while a further gene, designated as *fadJ*,

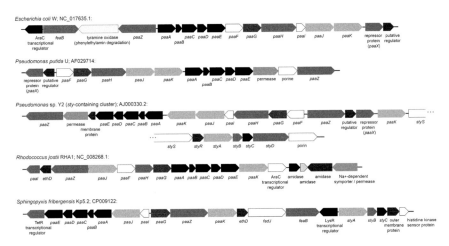

Fig. 3.3 Selected *paa*-gene clusters are compared. Here the *paa*-gene clusters of *Escherichia coli* W (Ferrández et al. 1998), *Pseudomonas putida* U (Luengo et al. 2001; Olivera et al. 1998), *Pseudomonas* sp. Y2 (Peso-Santos et al. 2006), *Rhodococcus jostii* RHA1 (Navarro-Llorens et al. 2005), and *Sphingopyxis fribergensis* Kp5.2 (Oelschlägel et al. 2015) are illustrated. Names of the proteins encoded by the genes based on O'Leary et al. (2002a) and Teufel et al. (2010): StyA oxygenase subunit of styrene monooxygenase; StyB reductase subunit of styrene monooxygenase; StyC styrene oxide isomerase; StyD or FeaB phenylacetaldehyde dehydrogenase; PaaK phenylacetate-CoA ligase; PaaABCDE ring 1,2-phenylacetyl-CoA epoxidase; PaaG ring 1,2-epoxyphenylacetyl-CoA isomerase; PaaZ oxepin-CoA hydrolase/3-oxo-5,6-dehydrosuberyl-CoA semialdehyde dehydrogenase; PaaJ 3-oxoadipyl-CoA thiolase/3-oxo-5,6-dehydrosuberyl-CoA thiolase; PaaF 2,3-dehydroadipyl-CoA hydratase; PaaH 3-hydroxyadipyl-CoA dehydrogenase; PaaI, thioesterase

was found. Further genes are involved in the regulation of the gene expression, but they can differ between the organisms and strains. An overview about different *paa*-gene clusters is given in Fig. 3.3. Abe-Yoshizumi et al. (2004) has previously shown for PaaK that these genes are subjects of a frequent gene transfer between bacteria which may be the reason for their relatively frequent occurrence in different species.

For a few organisms, a direct linkage of the styrene-degrading gene cluster and the phenylacetic acid degradation cluster has been reported. In *Sphingopyxis fribergensis* Kp5.2, a phenylacetaldehyde dehydrogenase encoded by *styD* (or otherwise designated as *feaB*) is directly embedded in the *paa* cluster and the genes *styA*, *styB*, *styC* (Fig. 3.1) as well as genes of some hypothetic or transcriptional relevant proteins are directly located upstream of the *paa* genes. A similar cluster also containing such *sty*-genes has previously been reported for *Pseudomonas* sp. strain Y2 (Peso-Santos et al. 2006).

3.5.2.1 FadJ—A Novel Multifunctional Protein of the Phenylacetic Acid Degradation?

In the case of *Sphingopyxis fribergensis* Kp5.2, a further enzyme annotated as FadJ (accession number: AJA07148) was identified in the *paa*-gene cluster which is encoded by a gene neighbored to *ethD* and downstream to the gene of the phenylacetaldehyde dehydrogenase FeaB. FadJ and the similar FadB are commonly involved in the anaerobic and aerobic β-oxidation of fatty acids (Campbell et al. 2003) and FadJ has recently been described for the first time in the context of phenylacetic acid degradation by Oelschlägel et al. (2015). In this current study, it was supposed that FadJ replaces PaaF and PaaH in this strain because it has also been shown by previous studies that FadJ and FadB are multifunctional enzymes harboring enoyl hydratase and 3-hydroxyacyl-CoA dehydrogenase activity (Campbell et al. 2003; Snell et al. 2002; Yang et al. 1988).

Alignment and functional annotation of conserved domains (by means of NCBI Conserved Domain Search, Marchler-Bauer and Bryant 2004; Marchler-Bauer et al. 2009, 2011, 2013) of the FadJ from the *paa*-gene cluster of strain Kp5.2 and of PaaF, PaaH, FadJ, and FadB from *E. coli* K-12 (Fig. 3.4) also strengthened the assumption of a functional similarity of the proteins mentioned. FadJ of strain Kp5.2 contains a crotonase/enoyl-CoA hydratase domain which is similar to the domains in FadB, FadJ, and PaaF of strain K-12. Additionally, three domains were found in FadJ of strain Kp5.2 which are responsible for 3-hydroxyacyl-CoA dehydrogenase activity in FadB, FadJ, and PaaH of strain K-12 (Fig. 3.4). Important amino acids, which are conserved in domains mentioned (determined by The UniProt Consortium 2014; He and Yang 1996; Yang and Elzinga 1993), were also identified in almost all cases or are substituted with a similar type of amino acid (Fig. 3.4). These results strongly indicate that FadJ, which is commonly involved in the fatty acid degradation of microorganisms (Campbell et al. 2003; Snell et al. 2002), harbors a PaaH and PaaF activity and replaces PaaH and

Alignment 1

Alignment 2

Identity of FadJ Kp5.2 (a-b) to FadJ Kp5.2 (a-b), FadB K-12, and PaaF K-12 (a-b): 51.5% (103/200 AA)

FadJ Kp5.2:
FadJ P77399.1 K-12:   Identity of FadJ K-12 (a-b) to FadJ Kp5.2 (a-b): 33.3% (67/198 AA)
FadB P21177.2 K-12:   Identity of FadB K-12 (a-b) to FadJ Kp5.2 (a-b): 32.2% (63/197 AA)
FadB P76082.1 K-12:
PaaF P76083.1 K-12:   Identity of PaaF K-12 (a-b) to FadJ Kp5.2 (a-b): 32.6% (63/193 AA)
PaaH P76083.1 K-12:

FadJ Kp5.2:
FadJ P77399.1 K-12:
FadB P21177.2 K-12:
FadB P76082.1 K-12:
PaaF P76083.1 K-12:
PaaH P76083.1 K-12:

FadJ Kp5.2:
FadJ P77399.1 K-12:   Identity of FadJ K-12 (c-d) to FadJ Kp5.2 (c-d): 58.6% (106/181 AA)
FadB P21177.2 K-12:
FadB P76082.1 K-12:
PaaF P76083.1 K-12:
PaaH P76083.1 K-12:

FadJ Kp5.2:   Identity of FadJ Kp5.2 (c-d) to FadJ K-12, FadB K-12, and PaaF K-12 (c-d): 51.5% (92/181 AA)
FadJ P77399.1 K-12:   Identity of FadB K-12 (c-d) to FadJ Kp5.2 (c-d): 57.8% (104/180 AA)
FadB P21177.2 K-12:
FadB P76082.1 K-12:
PaaF P76083.1 K-12:   Identity of PaaF K-12 (c-d) to FadJ Kp5.2 (c-d): 42.2% (76/180 AA)
PaaH P76083.1 K-12:

FadJ Kp5.2:   Identity of FadJ Kp5.2 (e-f) to FadJ K-12, FadB K-12, and PaaF K-12 (e-f): 51.5% (52/101 AA)
FadJ P77399.1 K-12:   Identity of FadJ K-12 (e-f) to FadJ Kp5.2 (e-f): 63.6% (35/68 AA)
FadB P21177.2 K-12:
FadB P76082.1 K-12:
PaaF P76083.1 K-12:   Identity of PaaF K-12 (e-f) to FadJ Kp5.2 (e-f): 34.7% (34/98 AA)
PaaH P76083.1 K-12:

FadJ Kp5.2:
FadJ P77399.1 K-12:   Identity of FadJ K-12 (g1-h) to FadJ Kp5.2 (g-h): 63.6% (21/68 AA)
FadB P21177.2 K-12:   Identity of FadB K-12 (g1-h) to FadJ Kp5.2 (g-h): 47.7% (22/97 AA)
FadB P76082.1 K-12:
PaaF P76082.1 K-12:   Identity of PaaH K-12 (g2-h) to FadJ Kp5.2 (g-h): 13.4% (11/82 AA)
PaaH P76083.1 K-12:

Legend:
crotonase/enoyl-coenzyme A (CoA) hydratase superfamily
3-hydroxyacyl-CoA dehydrogenase NAD-binding domain
3-hydroxyacyl-CoA dehydrogenase, C-terminal domain
2nd 3-hydroxyacyl-CoA dehydrogenase, C-terminal domain
E.M.... amino acids necessary for activity, substrate, or cofactor binding

◀ **Fig. 3.4** A comparison of FadJ from strain Kp5.2 with PaaF, PaaH, FadJ, and FadB from *Escherichia coli* K-12 is shown. The protein sequences of PaaF, PaaH, FadJ, and FadB were aligned by ClustalX version 1.8 (Higgins and Sharp 1988; Thompson et al. 1997) and Gene-Doc version 2.6.003 (Nicholas et al. 1997). NCBI Conserved Domain Search (Marchler-Bauer and Bryant 2004; Marchler-Bauer et al. 2009, 2011, 2013) was applied to identify functional domains. Identities of the domains compared to each other are illustrated, too. Amino acids which are conserved in domains or are necessary for the activity and substrate or cofactor binding were identified by UniProt (The UniProt Consortium 2014; He and Yang 1996; Yang and Elzinga 1993) and are also labeled

PaaF in strain Kp5.2, respectively. Because of the supposed function of the FadJ protein described in the study of Oelschlägel et al. (2015), we suggest the nomenclature PaaFH for the enzyme and *paaFH* for the corresponding gene. Based on these results it can be assumed that PaaH and PaaF evolutionarily originated from such multifunctional enzymes as FadB or FadJ which are incorporated in the fatty acid degradation.

3.5.2.2 Regulation of the Phenylacetic Acid Degradation

First investigations on the regulation of the *paa*-genes were performed for *P. putida* U (Olivera et al. 1998) and *E. coli* W (Ferrández et al. 1998). In both cases, a repressor protein designated as PaaX is important for the transcription process. Investigations of Ferrández et al. (1998) have shown that PaaX prevents the expression of the *paa*-genes in *E. coli* W until phenylacetic acid was added. Further experiments with the disrupted PaaX variant resulted in a constitutive expression of *paa*-genes as PaaK. Similar results were obtained in a study of Olivera et al. (1998) which investigated the *paa* regulation in *P. putida* U. Expression experiments with promoters of the *paa*-gene cluster and, especially, the *paaX* gene from *E. coli* W in another *E. coli* strain, which harbors no own *paa*-genes, allowed the identification of the true inducer of this pathway in strain Ferrández et al. (1998) have shown that the addition of phenylacetic acid leads only to an expression in presence PaaX if also PaaK is present. This strongly indicates that mainly phenylacetyl-CoA induces the *paa* genes and not the acid itself.

Olivera et al. (1998) also described a strong repression effect in *P. putida* U if glucose was added in a final concentration of 5 mM. This repression is also influenced by the PaaX protein because inactivation of *paaX* by mutation strongly reduces the catabolic repression of glucose. A study of Peso-Santos et al. (2006) also describes an interaction of the PaaX protein in the regulation of the *sty*-operon in *Pseudomonas* sp. Y2. PaaX could interact with the *styA* promotor (Fig. 3.2) and represses the transcription of *sty*-genes. The PaaX regulator functions as an opponent of the StyS/StyR inducing system. It has been supposed by the authors that PaaX modulates the StyS/StyR-induced expression of *sty*-genes until phenylacetyl-CoA is formed and the complete pathway is induced (Peso-Santos et al. 2006).

References

Abe-Yoshizumi R, Kamei U, Yamada A, Kimura M, Ichihara S (2004) The evolution of the phenylacetic acid degradation pathway in bacteria. Biosci Biotechnol Biochem 68:746–748

Arenghi FLG, Berlanda D, Galli E, Sello G, Barbieri P (2001) Organization and regulation of *meta* cleavage pathway genes for toluene and o-xylene derivative degradation in *Pseudomonas stutzeri* OX1. Appl Environ Microbiol 67:3304–3308

Arias S, Olivera ER, Arcos M, Naharro G, Luengo JM (2008) Genetic analyses and molecular characterization of the pathways involved in the conversion of 2-phenylethylamine and 2-phenylethanol into phenylacetic acid in *Pseudomonas putida* U. Environ Microbiol 10:413–432

Baikalov I, Schröder I, Kaczor-Grzeskowiak M, Grzeskowiak K, Gunsalus RP, Dickerson RE (1996) Structure of the *Escherichia coli* response regulator NarL. Biochemistry 35:11053–11061

Bartolomé-Martín D, Martínez-García E, Mascaraque V, Rubio J, Perera J, Alonso S (2004) Characterization of a second functional gene cluster for the catabolism of phenylacetic acid in *Pseudomonas* sp. strain Y2. Gene 341:167–179

Beltrametti F, Marconi AM, Bestetti G, Galli E, Ruzzi M, Zennaro E (1997) Sequencing and functional analysis of styrene catabolism genes from *Pseudomonas fluorescens* ST. Appl Environ Microbiol 63:2232–2239

Bestetti G, Galli E, Ruzzi M, Baldacci G, Zennaro E, Frontali L (1984) Molecular characterization of a plasmid from *Pseudomonas fluorescens* involved in styrene degradation. Plasmid 12:181–188

Bijlsma JJE, Groisman EA (2003) making informed decisions: regulatory interactions between two-component systems. Trends Microbiol 11:359–366

Braun-Lüllemann A, Majcherczyk A, Huttermann A (1997) Degradation of styrene by white-rot fungi. Appl Microbiol Biotechnol 47:150–155

Campbell JW, Morgan-Kiss RM, Cronan JE Jr (2003) A new *Escherichia coli* metabolic competency: growth on fatty acids by a novel anaerobic *beta*-oxidation pathway. Mol Microbiol 47:793–805

Chen X, Kohl TA, Rückert C, Rodionov DA, Li L-H, Ding J-Y, Kalinowski J, Liu S-J (2012) Phenylacetic acid catabolism and its transcriptional regulation in *Corynebacterium glutamicum*. Appl Environ Microbiol 78:5796–5804

Cho MC, Kang D-O, Yoon BD, Lee K (2000) Toluene degradation pathway from *Pseudomonas putida* F1: substrate specificity and gene induction by 1-substituted benzenes. J Ind Microbio Biotech 25:163–170

Coschigano PW, Young LY (1997) Identification and sequence analysis of two regulatory genes involved in anaerobic toluene metabolism by strain T1. Appl Environ Microbiol 63:652–660

del Peso-Santos T, Bartolomé-Martín D, Fernández C, Alonso S, García JL, Díaz E, Shingler V, Perera J (2006) Coregulation by phenylacetyl-Coenzyme A-responsive PaaX integrates control of the upper and lower pathways for catabolism of styrene by *Pseudomonas* sp. strain Y2. J Bacteriol 188:4812–4821

Di Gennaro P, Ferrara S, Ronco I, Galli E, Sello G, Papacchini M, Bestetti G (2007) Styrene lower catabolic pathway in *Pseudomonas fluorescens* ST: identification and characterization of genes for phenylacetic acid degradation. Arch Microbiol 188:117–125

Dobslaw D, Engesser K-H (2014) Degradation of toluene by *ortho* cleavage enzymes in *Burkholderia fungorum* FLU100. Microb Biotechnol 8:143–154

Ferrández A, Miñambres B, García B, Olivera ER, Luengo JM, García JL, Díaz E (1998) Catabolism of phenylacetic acid in *Escherichia coli*. Characterization of a new aerobic hybrid pathway. J Biol Chem 273:25974–25986

Gallegos MT, Schleif R, Bairoch A, Hofmann K, Ramos JL (1997) Arac/XylS family of transcriptional regulators. Microbiol Mol Biol Rev 61:393–410

Grebe TW, Stock JB (1999) The histidine protein kinase superfamily. Adv Microb Physiol 41:139–227

Gröning JAD, Kaschabek SR, Schlömann M, Tischler D (2014) A mechanistic study on SMOB-ADP1: an NADH:flavin oxidoreductase of the two-component styrene monooxygenase of *Acinetobacter baylyi* ADP1. Arch Microbiol 196:829–845

Hartmans S, Smits JP, van der Werf MJ, Volkering F, de Bont JAM (1989) Metabolism of styrene oxide and 2-phenylethanol in the styrene-degrading *Xanthobacter* strain 124X. Appl Environ Microbiol 55:2850–2855

He XY, Yang SY (1996) Histidine-450 is the catalytic residue of L-3-hydroxyacyl coenzyme A dehydrogenase associated with the large alpha-subunit of the multienzyme complex of fatty acid oxidation from *Escherichia coli*. Biochemistry 35:9625–9630

Higgins DG, Sharp PM (1988) CLUSTAL: a package for performing multiple sequence alignment on a microcomputer. Gene 73:237–244

Huijbers MME, Montersino S, Westphal AH, Tischler D, van Berkel WJH (2014) Flavin dependent monooxygenases. Arch Biochem Biophys 544:2–17

Itoh N, Hayashi K, Okada K, Ito T, Mizuguchi N (1997) Characterization of styrene oxide isomerase, a key enzyme of styrene and styrene oxide metabolism in *Corynebacterium* sp. Biosci Biotech Biochem 61:2058–2062

Kantz A, Chin F, Nallamothu N, Nguyen T, Gassner GT (2005) Mechanism of flavin transfer and oxygen activation by the two-component flavoenzyme styrene monooxygenase. Arch Biochem Biophys 442:102–116

Köhler KAK, Rückert C, Schatschneider S, Vorhölter F-J, Szczepanowski R, Blank LM, Niehaus K, Goesmann A, Pühler A, Kalinowski J, Schmid A (2013) Complete genome sequence of *Pseudomonas* sp. strain VLB120 a solvent tolerant, styrene degrading bacterium, isolated from forest soil. J Biotechnol 168:729–730

Kukor JJ, Olsen RH (1991) Genetic organization and regulation of a *meta* cleavage pathway for catechols produced from catabolism of toluene, benzene, phenol, and cresols by *Pseudomonas pickettii* PKO1. J Bacteriol 173:4587–4594

Labbé D, Garnon J, Lau PC (1997) Characterization of the genes encoding a receptor-like histidine kinase and a cognate response regulator from a biphenyl/polychlorobiphenyl-degrading bacterium, *Rhodococcus* sp. strain M5. J Bacteriol 179:2772–2776

Lau PC, Wang Y, Patel A, Labbé D, Bergeron H, Brousseau R, Konishi Y, Rawlings M (1997) A bacterial basic region leucine zipper histidine kinase regulating toluene degradation. Proc Natl Acad Sci USA 94:1453–1458

Leoni L, Ascenzi P, Bocedi A, Rampioni G, Castellini L, Zennaro E (2003) Styrene-catabolism regulation in *Pseudomonas fluorescens* ST: phosphorylation of StyR induces dimerization and cooperative DNA-binding. Biochem Biophys Res Commun 303:926–931

Leoni L, Rampioni G, Stefano VD, Zennaro E (2005) Dual role of response regulator StyR in styrene catabolism regulation. Appl Environ Microbiol 71:5411–5419

Leuthner B, Heider J (1998) A two-component system involved in regulation of anaerobic toluene metabolism in *Thauera aromatica*. FEMS Microbiol Lett 166:35–41

Lin H, Qiao J, Liu Y, Wu Z-L (2010) Styrene monooxygenase from *Pseudomonas* sp. LQ26 catalyzes the asymmetric epoxidation of both conjugated and unconjugated alkenes. J Mol Catal B Enzym 67:236–241

Luengo JM, García JL, Olivera ER (2001) The phenylacetyl-CoA catabolon: a complex catabolic unit with broad biotechnological applications. Mol Microbiol 39:1343–1442

Maltseva OV, Solyanikova IP, Golovleva LA, Schlömann M, Knackmuss HJ (1994) Dienelactone hydrolase from *Rhodococcus erythropolis* 1CP: purification and properties. Arch Mircobiol 162:368–374

Marchler-Bauer A, Bryant SH (2004) CD-Search: protein domain annotations on the fly. Nucleic Acids Res 32:W327–W331

Marchler-Bauer A, Anderson JB, Chitsaz F, Derbyshire MK, DeWeese-Scott C, Fong JH, Geer
    LY, Geer RC, Gonzales NR, Gwadz M, He S, Hurwitz DI, Jackson JD, Ke Z, Lanczycki CJ,
    Liebert CA, Liu C, Lu F, Lu S, Marchler GH, Mullokandov M, Song JS, Tasneem A, Thanki
    N, Yamashita RA, Zhang D, Zhang N, Bryant SH (2009) CDD: specific functional annotation
    with the conserved domain database. Nucleic Acids Res 37:D205–D210
Marchler-Bauer A, Lu S, Anderson JB, Chitsaz F, Derbyshire MK, DeWeese-Scott C, Fong JH,
    Geer LY, Geer RC, Gonzales NR, Gwadz M, Hurwitz DI, Jackson JD, Ke Z, Lanczycki CJ,
    Lu F, Marchler GH, Mullokandov M, Omelchenko MV, Robertson CL, Song JS, Thanki N,
    Yamashita RA, Zhang D, Zhang N, Zheng C, Bryant SH (2011) CDD: a Conserved Domain
    Database for the functional annotation of proteins. Nucleic Acids Res 39:D225–D229
Marchler-Bauer A, Zheng C, Chitsaz F, Derbyshire MK, Geer LY, Geer RC, Gonzales NR,
    Gwadz M, Hurwitz DI, Lanczycki CJ, Lu F, Lu S, Marchler JS, Song GH, Thanki N,
    Yamashita RA, Zhang D, Bryant SH (2013) CDD: conserved domains and protein three-
    dimensional structure. Nucleic Acids Res 41:D348–D352
Marconi AM, Beltrametti F, Bestetti G, Solinas F, Ruzzi M, Galli E, Zennaro E (1996) Cloning
    and characterization of styrene catabolism genes from Pseudomonas fluorescens ST. Appl
    Environ Microbiol 62:121–127
Massai F, Rampioni G, Micolonghi C, Messina M, Zennaro E, Ascenzi P, Leoni L (2014) Styrene
    is sensed by the N-terminal PAS sensor domain of StyS, a double sensor kinase from the sty-
    rene-degrading bacterium Pseudomonas fluorescens ST. Ann Microniol Epub ahead of print
Milani M, Leoni L, Rampioni G, Zennaro E, Ascenzi P, Bolognesi M (2005) An active-like struc-
    ture in the unphosphorylated StyR response regulator suggests a phosphorylation-dependent
    allosteric activation mechanism. Structure 13:1289–1297
Montersino S, Tischler D, Gassner GT, van Berkel WJH (2011) Catalytic and structural features
    of flavoprotein hydroxylases and epoxidases. Adv Synth Catal 353:2301–2319
Mooney A, O'Leary ND, Dobson ADW (2006) Cloning and functional characterization of
    the styE gene, involved in styrene transport in Pseudomonas putida CA-3. Appl Environ
    Microbiol 72:1302–1309
Mosqueda G, Ramos JL (2000) A set of genes encoding a second toluene efflux system in
    Pseudomonas putida DOT-T1E is linked to the tod genes for toluene metabolism. J Bacteriol
    182:937–943
Navarro-Llorens JM, Patrauchan MA, Stewart GR, Davies JE, Eltis LD, Mohn WW (2005)
    Phenylacetate catabolism in Rhodococcus sp. strain RHA1: a central pathway for degradation
    of aromatic compounds. J Bacteriol 187:4497–4504
Nicholas KB, Nicholas HBJ, Deerfield DWI (1997) GeneDoc: analysis and visualization of
    genetic variation. Embnew News 4:14
Nikodinovic-Runic J, Flanagan M, Hume AR, Cagney G, O'Connor KE (2009) Analysis of the
    Pseudomonas putida CA-3 proteome during growth on styrene under nitrogen-limiting and
    non-limiting conditions. Microbiology 155:3348–3361
O'Connor K, Buckley CM, Hartmans S, Dobson AD (1995) Possible regulatory role for non-
    aromatic carbon sources in styrene degradation by Pseudomonas putida CA-3. Appl Environ
    Microbiol 61:544–548
O'Leary ND, Duetz WA, Dobson ADW, O'Connor KE (2002a) Induction and repression of
    the sty operon in Pseudomonas putida CA-3 during growth on phenylacetic acid under
    organic and inorganic nutrient-limiting continuous culture conditions. FEMS Microbiol Lett
    208:263–268
O'Connor KE, Dobson AD, Hartmans S (1997) Indigo formation by microorganisms expressing
    styrene monooxygenase activity. Appl Environ Microbiol 63:4287–4291
Oelschlägel M, Gröning JAD, Tischler D, Kaschabek SR, Schlömann M (2012) Styrene oxide
    isomerase of Rhodococcus opacus 1CP, a highly stable and considerably active enzyme. Appl
    Environ Microbiol 78:4330–4337
Oelschlägel M, Zimmerling J, Schlömann M, Tischler D (2014) Styrene oxide isomerase of
    Sphingopyxis sp. Kp5.2. Microbiology (UK) 160:2481–2491

Oelschlägel M, Rückert C, Kalinowski J, Schmidt G, Schlömann M, Tischler D (2015) Description of *Sphingopyxis fribergensis* sp. nov.—a soil bacterium with the ability to degrade styrene and phenylacetic acid. Int J Syst Evol Microbiol. doi:10.1099/ijs.0.000371

O'Leary ND, O'Connor KE, Duetz W, Dobson ADW (2001) Transcriptional regulation of styrene degradation in *Pseudomonas putida* CA-3. Microbiology 147:973–979

O'Leary ND, O'Connor KE, Dobson ADW (2002b) Biochemistry, genetics and physiology of microbial styrene degradation. FEMS Microbiol Rev 26:403–417

Olivera ER, Miñambres B, García B, Muñiz C, Moreno MA, Ferrández A, Díaz E, García JL, Luengo JM (1998) Molecular characterization of the phenylacetic acid catabolic pathway in *Pseudomonas putida* U: the phenylacetyl-CoA catabolon. Proc Natl Acad Sci USA 95:6419–6424

Otto K, Hofstetter K, Roethlisberger M, Witholt B, Schmid A (2004) Biochemical characterization of StyAB from *Pseudomonas* sp. strain VLB120 as a two-component flavin-diffusible monooxygenase. J Bacteriol 186:5292–5302

Panke S, Witholt B, Schmid A, Wubbolts MG (1998) Towards a biocatalyst for (*S*)-styrene oxide production: characterization of the styrene degradation pathway of *Pseudomonas* sp. strain VLB120. Appl Environ Microbiol 64:2032–2043

Parkinson JS, Kofoid EC (1992) Communication modules in bacterial signaling proteins. Annu Rev Genet 26:71–112

Patrauchan MA, Florizone C, Eapen S, Gómez-Gil L, Sethuraman B, Fukuda M, Davies J, Mohn WW, Eltis LD (2008) Roles of ring-hydroxylating dioxygen-ases in styrene and benzene catabolism in *Rhodococcus jostii* RHA1. J Bacteriol 190:37–47

Reizer J, Saier MH (1997) Modular multidomain phosphoryl transfer proteins of bacteria. Curr Opin Struct Biol 7:407–415

Ruzzi M, Zennaro E (1989) pEG plasmid involved in styrene degradation: molecular dimorphism and integration of a segment into the chromosome. FEMS Microbiol Lett 50:337–343

Santos PM, Blatny JM, Bartolo ID, Valla S, Zennaro E (2000) Physiological analysis of the expression of the styrene degradation gene cluster in *Pseudomonas fluorescens* ST. Appl Environ Microbiol 66:1305–1310

Santos PM, Leoni L, Bartolo ID, Zennaro E (2002) Integration host factor is essential for the optimal expression of the *styABCD* operon in *Pseudomonas fluorescens* ST. Res Microbiol 153:527–536

Shirai K, Hisatsuka K (1979) Production of β-phenethyl alcohol from styrene by *Pseudomonas* 305-STR-1-4. Agric Biol Chem 43:1399–1406

Snell KD, Feng F, Zhong L, Martin D, Madison LL (2002) YfcX enables medium-chain-length poly(3-hydroxyalkanoate) formation from fatty acids in recombinant *Escherichia coli* fadB strains. J Bacteriol 184:5696–5705

Taylor BL, Zhulin IB (1999) PAS domains: internal sensors of oxygen, redox potential, and light. Microbiol Mol Biol Rev 63:479–506

Teufel R, Mascaraque V, Ismail W, Voss M, Perera J, Eisenreich W, Haehnel W, Fuchs G (2010) Bacterial phenylalanine and phenylacetate catabolic pathway revealed. Proc Natl Acad Sci USA 107:14390–14395

The Uniprot Consortium (2014) Activities at the Universal Protein Resource (UniProt). Nucleic Acids Res 42:D191–D198

Thompson JD, Gibson TJ, Plewniak F, Jeanmougin F, Higgins DG (1997) The CLUSTAL_X windows interface: flexible strategies for multiple sequence alignment aided by quality analysis tools. Nucleic Acids Res 25:4876–4882

Tischler D, Kaschabek SR (2012) Microbial degradation of xenobiotics. In: Singh SN (ed). Springer, Berlin, pp 67–99

Tischler D, Eulberg D, Lakner S, Kaschabek SR, van Berkel WJH, Schlömann M (2009) Identification of a novel self-sufficient styrene monooxygenase from *Rhodococcus opacus* 1CP. J Bacteriol 191:4996–5009

Tischler D, Kermer R, Gröning JAD, Kaschabek SR, van Berkel WJH, Schlömann M (2010) StyA1 and StyA2B from *Rhodococcus opacus* 1CP: a multifunctional styrene monooxygenase system. J Bacteriol 192:5220–5227

Tischler D, Gröning JAD, Kaschabek SR, Schlömann M (2012) One-component styrene monooxygenases: an evolutionary view on a rare class of flavoproteins. Appl Biochem Biotechnol 167:931–944

Tischler D, Schlömann M, van Berkel WJH, Gassner GT (2013) FAD C(4a)-hydroxide stabilized in a naturally fused styrene monooxygenase. FEBS Lett 587:3848–3852

Toda H, Itoh N (2012) Isolation and characterization of styrene metabolism genes from styrene-assimilating soil bacteria *Rhodococcus* sp. ST-5 and ST-10. J Biosci Bioeng 113:12–19

Utkin I, Yakimov M, Matveeva L, Kozlyak E, Rogozhin I, Solomon Z, Bezborodov A (1991) Degradation of styrene and ethylbenzene by *Pseudomonas* species Y2. FEMS Microbiol Lett 77:237–242

van Hellemond EW, Janssen DB, Fraaije MW (2007) Discovery of a novel styrene monooxygenase originating from the metagenome. Appl Environ Microbiol 73:5832–5839

Velasco A, Alonso S, Garcia JL, Perera J, Diaz E (1998) Genetic and functional analysis of the styrene catabolic cluster of *Pseudomonas* sp. strain Y2. J Bacteriol 180:1063–1071

Volmer J, Neumann C, Bühler B, Schmid A (2014) Engineering of *Pseudomonas taiwanensis* VLB120 for constitutive solvent tolerance and increased specific styrene epoxidation activity. Appl Environ Microbiol 80:6539–6548

Warhurst AM, Clarke KF, Hill RA, Holt RA, Fewson CA (1994) Metabolism of styrene by *Rhodococcus rhodochrous* NCIMB 13259. Appl Environ Microbiol 60:1137–1145

Werlen C, Kohler H-PE, van der Meer JR (1996) The broad substrate chlorobenzene dioxygenase and cis-chlorobenzene dihydrodiol dehydrogenase of *Pseudomonas* sp. Strain P51 are linked evolutionarily to the enzymes for benzene and toluene degradation. J Biol Chem 271:4009–4016

Yang SY, Elzinga M (1993) Association of both enoyl coenzyme A hydratase and 3-hydroxyacyl coenzyme A epimerase with an active site in the amino-terminal domain of the multifunctional fatty acid oxidation protein from *Escherichia coli*. J Biol Chem 268:6588–6592

Yang SY, Li JM, He XY, Cosloy SD, Schulz H (1988) Evidence that the *fadB* gene of the *fadAB* operon of *Escherichia coli* encodes 3-hydroxyacyl-coenzyme A (CoA) epimerase, delta 3-cis-delta 2-trans-enoyl-CoA isomerase, and enoyl-CoA hydratase in addition to 3-hydroxyacyl-CoA dehydrogenase. J Bacteriol 170:2543–2548

Chapter 4
Selected Enzymes of Styrene Catabolism

Abstract During the last decades attention was drawn to several enzymes of styrene catabolic routes with respect to general behavior and activity, mechanism, structure, and biotechnological applications. Especially, various styrene monooxygenases from different bacteria were investigated. These enzymes are of special interest since they allow the initial activation of styrene and therewith its breakdown. Further, they produce enantiopure epoxides and thus interesting building blocks for fine chemical synthesis. But, also enzymes acting on further intermediates as styrene oxide and phenylacetaldehyde have been highlighted. Styrene oxide isomerase performs the cofactor independent isomerization of the styrene epoxide to the phenylacetaldehyde. The latter is a valuable flavor compound, thus these enzymes were studied with respect to their application in aldehyde production. The respective phenylacetaldehyde can either be reduced by a reductase yielding 2-phenylethanol or oxidized by a dehydrogenase yielding phenylacetic acid. Epoxide hydrolase is a representative of a fungal styrene metabolic activity and thus another relevant enzyme acting on styrene oxide. It allows the selective resolution of racemic epoxides to yield a diol and enantiopure epoxide. Herein, a brief description of mentioned enzymes activating and converting styrene and its metabolites within styrene degradation routes is provided.

Keywords Styrene catabolic enzymes · Epoxidase · Epoxide isomerase · Epoxide hydrolase · Phenylacetaldehyde

4.1 Styrene Monooxygenase

So far, class E flavoprotein monooxygenases are solely represented by styrene monooxygenases (SMO; EC 1.14.14.11) (Huijbers et al. 2014). SMOs convert regio- and enantioselective styrene to (S)-styrene oxide and had first been described

© The Author(s) 2015
D. Tischler, *Microbial Styrene Degradation*,
SpringerBriefs in Microbiology, DOI 10.1007/978-3-319-24862-2_4

by Hartmans et al. (1990) as flavin adenine dinucleotide (FAD)-dependent enzymes. These studies attended with the capability of some microorganisms to utilize styrene as a sole source of carbon and energy (see Chaps. 2 and 3). In the meantime, many more investigations revealed SMOs from various microorganisms or (meta)genomes (Beltrametti et al. 1997; Guan et al. 2007; Lin et al. 2010; Marconi et al. 1996; O'Connor et al. 1997; Oelschlägel et al. 2014a, b; Panke et al. 1998; Park et al. 2006; Tischler et al. 2009, 2012; Toda et al. 2012; van Hellemond et al. 2007; Velasco et al. 1998). These enzymes have been studied for their physiological role as initial biocatalysts of the styrene mineralization process (reviewed: O'Leary et al. 2002; Mooney et al. 2006; Tischler and Kaschabek 2012). But, lately more often biochemical properties and structural aspects have been investigated and frequently these enzymes are evaluated as potential biocatalysts producing enantiopure building blocks for fine chemical syntheses (see Chap. 5).

4.1.1 The Two Types of Styrene Monooxygenases

In general, SMOs are two-component enzymes composed of an NADH-utilizing flavin reductase (StyB) and a styrene epoxidizing monooxygenase (StyA) (Figs. 4.1 and 4.3) (Montersino et al. 2011). Reductases use solely NADH to

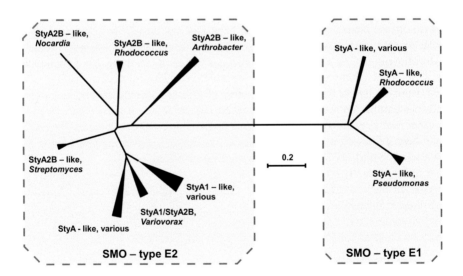

Fig. 4.1 The phylogenetic distance tree of SMOs allows distinguishing two types of monooxygenases. The phylogenetic distance tree was calculated as earlier shown (Tischler et al. 2012), but now with more sequences available from the database/literature. Clustering of both SMO types (E1 and E2) gets obvious and is indicated by gray boxing. SMO proteins with high homology were sub-clustered as indicated. Interestingly, the StyA1- and StyA2B-like proteins of *Variovorax* species form a cluster next to other StyA- or StyA1-like proteins of the E2 type indicating a shorter evolutionary history (Tischler et al. 2012)

reduce flavins, while NADPH does not serve as electron donor. But, the electron acceptor can be riboflavin, flavin mononucleotide (FMN), and FAD. Interestingly, the epoxidases just accept reduced FAD to perform regio- and enantioselective monooxygenation of styrene by first activating and then using molecular oxygen as the reactant. Typically, for class E flavoprotein monooxygenases is that the reductase activity exceeds that of the epoxidase (Otto et al. 2004; Tischler et al. 2009, 2010; Toda et al. 2012) and thus an overproduction of reduced FAD gets apparent. Free reduced FAD reacts rapidly with molecular oxygen and leads to the unproductive formation of reactive oxygen as hydrogen peroxide or superoxide (Massey 1994).

The first more in detail described SMOs have been discovered among pseudomonads (Beltrametti et al. 1997; Marconi et al. 1996; Panke et al. 1998; Velasco et al. 1998). And here, the prototype SMO is represented by StyA and StyB originating from *Pseudomonas putida* S12 or *Pseudomonas taiwanensis* VLB120 (Fig. 4.1) (Kantz et al. 2005; Otto et al. 2004; Vollmer et al. 2014) for which the subunits are identical on amino acid level (StyA and StyB of both strains show 100 % identical positions over a length of 415 and 170 amino acids, respectively). These enzymes were later subject for a variety of kinetic, structural, as well as biotechnological studies. Respectively, it is a typical SMO as described above consisting of a single reductase (StyB) and a single epoxidase (StyA) and can be considered as the conventional SMO type E1 (Montersino et al. 2011). The SMO is part of a styrene catabolic gene cluster (*sty*-cluster; see Sect. 3.1) and initiates the degradation of styrene by activating it to the corresponding epoxide which undergoes further oxidation toward phenylacetic acid as a key metabolite (O'Leary et al. 2002; Mooney et al. 2006; Tischler and Kaschabek 2012). Up to now several similar and active SMOs (7 from pseudomonads, 3 from rhodococci, and 3 from other sources) of this type were identified and investigated in more detail. Respectively, these are mostly part of the mentioned *sty*-cluster. Besides those also a large number of putative monooxygenases were found and annotated as SMOs (Tischler et al. 2012).

More recently, a second type of SMO (E2) has been discovered and has initially been described in dependence on the E1-SMO of strain VLB120 (Otto et al. 2004; Tischler et al. 2009; van Hellemond et al. 2007). The respective prototype SMO is represented by StyA1 and StyA2B from *Rhodococcus opacus* 1CP (Fig. 4.1) (Tischler et al. 2009, 2010). In this case, the reductase is naturally fused to a second epoxidase subunit (StyA2-fused to B), making this the first active self-sufficient SMO reported. Indeed, it was proven that the reductase and the fused epoxidase allow the epoxidation of styrene by following the reaction pattern described above. However, the epoxidase power was found to be rather low (0.02 U mg^{-1}). And as with the E1 SMO type, here an additional single major epoxidase (StyA1) belongs to StyA2B, whereas the latter acts mainly as the reductase of that system. The reducase part of StyA2B serves efficiently as a source of reducing equivalents for both epoxidase subunits (StyA1 and StyA2-fused). Together, StyA1 and StyA2B perform an efficient and comparable epoxidation reaction with about 0.24 U mg^{-1} (Tischler et al. 2010) as the conventional type SMOs with 0.025–2.1 U mg^{-1} (Otto et al. 2004; Riedel et al. 2015; Toda et al.

2012; van Hellemond et al. 2007). However, it needs to be mentioned that recombinant StyA1 in combination with a chemical reductant achieves highest catalytic efficiency (Paul et al. 2015).

So far, the E2 type of SMOs, which is composed of a single epoxidase and the fused epoxidase-reductase component, was found only in a few bacteria (5 from *Rhodococcus*, 21 from *Streptomyces*, 1 from *Nocardia*, 9 from *Arthrobacter*, 1 from *Sciscionella*, 1 from *Amycolatopsis*, and 6 from *Variovorax*) and so far only a single system has been studied in detail (StyA1 and StyA2B of *R. opacus* 1CP) (Paul et al. 2015; Riedel et al. 2015; Tischler et al. 2009, 2010, 2013). Most of them belong to the GC-rich Gram-positive Actinobacteria and have only been annotated by bioinformatic tools. Any proof of functionality need to be demonstrated. Interestingly, these SMOs are not encoded within a styrene catabolic gene cluster, but in another one with, so far, no physiological assignment (Tischler et al. 2009, 2012). Furthermore, it hast to be mentioned that a phylogenetic analysis showed that this type of SMO is distinct from the conventional E1 SMO type (Fig. 4.1). But, it has also been revealed that this SMO type is more diverse in respect of protein subunit organization since also homologous SMO proteins of the StyA/StyB organization were found to be closely related to the StyA1/StyA2B type. Thus we suggest not only protein organization, but also the phylogenetic affiliation needs to be addressed to differentiate both types of SMOs (E1 and E2). Mechanistic and structural aspects may even be different as well and could confirm a certain annotation.

4.1.2 Insights into Structure and Mechanism of SMOs

4.1.2.1 StyA and StyB from *P. putida* S12

The only experimental structure information of SMOs is provided for StyA and StyB from *P. putida* S12, so far (Morrison et al. 2013; Ukaegbu et al. 2010). Further in silico studies on the protein structure and binding of substrates to StyA were performed and in few cases by mutagenesis analyzed (Feenstra et al. 2006; Gursky et al. 2010; Lin et al. 2012; Qaed et al. 2011). But, no structural information on the SMO E2 type was reported yet. However, in case of StyA1, a relative similar fold as for StyA can be expected due to the high sequence similarity reported (Tischler et al. 2009). And for StyA2B, the individual subunits might also have structural similarity to StyA and StyB. The natural fusion protein is missing about 10 amino acids at the N-terminal StyB-subunit part in comparison to StyB-like proteins and therefore the linker-region and its three-dimensional structure remain enigmatic.

The three-dimensional structure of StyA (PDB: 3IHM) has already been described and reviewed (Huijbers et al. 2014; Montersino et al. 2011; Ukaegbu et al. 2010) in respect to substrate binding. Furthermore, general elements of the fold in comparison to 4-hydroxybenzoate 3-hydroxylase (PHBH; PDB: 1cc6) were identified and found to be conserved (Huijbers et al. 2014; Riedel et al. 2015).

Differences as the proposed NAD(P)H-binding site which is present in PHBH and substrate pocket were highlighted and discussed. Thus, here we omit a detailed structural report while referring to respective reviews (Huijbers et al. 2014; Montersino et al. 2011; van Berkel et al. 2006). Nevertheless, a recent study based on in silico styrene-docking into StyA, sequence alignments, and afterward experimental verification has to be mentioned (Lin et al. 2012). Aiming to identify crucial residues for styrene binding and therewith activity in silico substrate docking had been performed based on the StyA structure (PDB: 3IHM). Three residues had been identified (Tyr73, His76, and Ser96). Subsequently, it has been confirmed by mutagenesis and activity measurements that Tyr73Val and Ser96Ala have indeed an effect on substrate turnover. Especially, the mutant Ser96Ala allowed efficient and faster substrate conversion by maintaining stereoselectivity. Interestingly, the Tyr73Val mutant converted 1-phenylcyclohexene into the (R,R)- and not into the (S,S)-enantiomer as the wild-type. That might be an indication that Tyr73 is involved in styrene binding next to the vinyl side chain effecting stereoselectivity. Respective mutations were found to be naturally present in E2-type SMOs. And those might therefore be interesting candidates to study stereoselectivity by site-directed mutagenesis.

The three-dimensional structure of StyB from *P. putida* S12 was uncovered very recently (Morrison et al. 2013). In analogy to earlier sequence alignments (Tischler et al. 2009), the StyB structure shows high homology to reductase structures of other two-component flavoprotein monooxygenases as exemplary PheA2 (PDB: 1RZ1), HpaC (PDB: 2ECR), and a putative flavin reductase (PDB: 1USF). The asymmetric unit counts 12 StyB-monomers and FAD molecules (PDB: 4F07). StyB was found to occur as a homodimeric protein containing a two-fold symmetry center. Each monomer binds two FAD molecules in an opposed direction while sharing them with another monomer (Fig. 4.2). Interestingly, one FAD binds similarly as observed in PheA2 while pointing with the isoalloxazine ring into the active site and having the adenine part showing out to the solvent in a more accessible form. Thus, this adenine could get close to the NADH binding site of another StyB monomer. Some differences in the loop formed by residues 94–102 of StyB compared to PheA2 (residues 84–92) indicated a weaker binding of FAD to StyB which is consistent with respective FAD dissociation constants (StyB: $K_d = 1.2$ μM, Morrison et al. 2013; and PheA2: $K_d = 0.01$ μM, van den Heuvel et al. 2004). The second FAD is binding vice versa to the first monomer, respectively, and thus interacting with the second monomer as well (Fig. 4.2). The NADH binding site and its binding seems to be similar for both reductases as well. However, a nitrate occupies the position of the proposed site for the nicotinamide moiety of NAD^+ shown in PheA2.

4.1.2.2 Mechanism of Conventional SMOs (StyA/StyB)

The reductases are strictly dependent on NADH as source of the reducing power and reduce various flavins (Otto et al. 2004; Kantz et al. 2005; Toda et al. 2012). A sequential ordered mechanism is reported for the apo-protein at which NADH

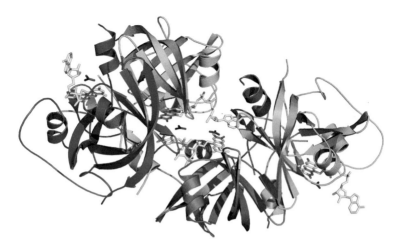

Fig. 4.2 The three-dimensional structure of StyB (PDB: 4F07; Morrison et al. 2013) represents a typical reductase-like fold. Here, four monomers of the asymmetric unit are shown which account for two dimers (*marine/turquoise* and *green/dark green*). The FAD binding (*yellow*) and thus interaction of two monomers (*turquoise* and *dark green*) is highlighted in the center. Each monomer binds two FAD molecules, one with the adenine and the other with the isoalloxazine ring and the rest of these FAD molecules point into the active site of the other StyB monomer. Nitrate (*red*) is shown at the position at which the nicotinamide of NAD$^+$ might be located

is the leading substrate. However, a tight binding of oxidized FAD to apoStyB was observed, and especially at high concentrations the NADH binding can so be blocked and catalysis be hindered (Morrison et al. 2013; Otto et al. 2004).

The second component StyA accepts solely reduced FAD for styrene epoxidation. No stable complex of StyB and StyA were observed which would indicate a direct transfer of FAD between subunits. But, the transfer of this reducing agent to the StyA active site is still object of investigation. Initial studies demonstrated a diffusive transport is possible (Hollman et al. 2003; Otto et al. 2004; Kantz et al. 2005) since when having a membrane between subunits fractions, or supplying chemical reduced FAD toward StyA a high steady-state epoxidation rate was still possible. However, already results obtained from these experiments raised questions since just diffusion allowed not to explain the complex StyB/StyA system. First indications of a transient complex during catalysis were obtained (Kantz et al. 2005). Later, mechanistic and structural evidence was provided thus both subunits may interact during catalysis (Morrison et al. 2013). StyA seems to stabilize apoStyB and elongate its lifetime. Furthermore, it was found that StyB holds on to reduced FAD which is gated by the releasing NAD$^+$ until it is transferred to StyA. And most striking in the presence of StyB StyA performed a faster epoxidation as if chemically reduced FAD was provided. Thus a protein–protein interaction seems to accelerate the system and makes it more efficient as well. The latter also became evident since the competitive use of reduced FAD from StyB by StyA versus cytochrome c did not end up just in cytochrome c reduction. The presence

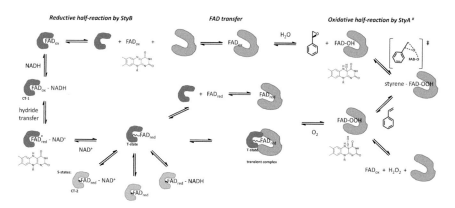

Fig. 4.3 Mechanism of StyB-mediated epoxidation of styrene by StyA (Kantz et al. 2005; Kantz and Gassner 2011; Morrison et al. 2013). During the reductive half-reaction oxidized FAD (FAD_{ox}) gets reduced by NADH in active site of StyB while first a charge transfer complex (CT-1) from NADH to FAD_{ox} is formed followed by a rapid hydride transfer yielding the reduced FAD (FAD_{red}). FAD_{red} bound to StyB can occur in an exposed (T) or sequestered (S) state in dependency of the molecules around. In the S-states, the reduced FAD seems to be stored or protected and in the T-state transferred to StyA for epoxidation. Here, a second charge transfer complex (CT-2) from FAD_{red} to NAD^+ can occur. The FAD_{red} can be released and then by diffusion transferred or as recently suggested by StyB–StyA interaction within a transient complex handed over. All further reactions of the oxidative site can thereby occur by StyA or StyB–StyA complex (not shown). However, FAD_{red} rapidly reacts with molecular oxygen to form a peroxy intermediate which either decomposes via uncoupling or allows after styrene binding epoxidation. Thereof, styrene oxide and hydroxy-FAD are formed while the epoxide gets released and the hydroxy-FAD is dehydrated to FAD_{ox} which then allows another cycle of reactions

of StyA significantly limited the reduction of cytochrome c indicating first that reduced FAD binds tightly to the epoxidase subunit and second it was buried or protected to some extent by StyB/StyA.

The epoxidation itself needs reduced FAD in the active site of StyA, molecular oxygen, and styrene. Molecular oxygen is first activated by protein bound and reduced FAD to form a (hydro)peroxy intermediate which is significantly stabilized by StyA (Kantz and Gassner 2011). The hydroperoxy-FAD interacts with the vinyl side chain of styrene to form a transient and activated complex with chiral information introduced by the active site residues of StyA and probably by the isoalloxazine system itself. This funnels into a rapid and stereoselective epoxidation of styrene (104 s^{-1}) and the formation of a highly fluorescent hydroxyl-FAD intermediate which immediately undergoes dehydration to form oxidized FAD and water ($0.82–1.4$ s^{-1}) (Kantz et al. Kantz and Gassner 2011; Morrison et al. 2013). Thus, the FAD obtained can be channeled back to the reductase for another reaction cycle of StyB/StyA.

The overall reaction scheme of StyB/StyA biocatalysis is presented in Fig. 4.3 and highlights all reaction intermediates and so far suggested subunit states.

4.1.2.3 First Mechanistic Insights for SMO Type E2

The prototype system is composed of a single epoxidase StyA1 and a naturally fused protein of epoxidase and reductase StyA2B, whereas the latter mainly serves as FAD reductase (Tischler et al. 2009). StyA1 represents the major epoxidase of the system since it has about 10-times higher epoxidation power in comparison to the A2-domain of the fusion protein StyA2B. Furthermore, it was shown that StyA1 accepts reduced FAD also from distinct sources as other flavin reductases or chemically reduced (Paul et al. 2015; Tischler et al. 2010, 2011, 2013). However, a protein–protein interaction of StyA1 and StyA2B was suggested occurring during catalysis since highest efficiency and epoxidation power was observed at a 1–1 ratio of respective proteins (Tischler et al. 2010). That would be in agreement with the E1-type SMOs for which also strong indications for protein cross-talk were reported (Morrison et al. 2013). Very recently, a higher efficiency with StyA1 in combination with a chemical reductant for FAD (1-benzyl-1,4-dihydronicotinamide) was observed (Paul et al. 2015). But, it need to be mentioned that for a 90 % electron transfer yield from the chemical reductant toward product a very high concentration of the reductant had to be supplied. However, the FAD concentration was in a similar range as for enzymatic systems. Thus this novel non-natural system needs to be investigated from a mechanistical point of view in future studies.

The reductase subunit of StyA2B accepts in principle the same flavin cosubstrates as StyB and is also strictly dependent on NADH (Tischler et al. 2009). Furthermore, a sequential binding mechanism for NADH/FAD of the reductase can be assumed according to observed results (Tischler et al. 2011, 2013). The reductase subunit acts independently of the epoxidase since it reduces flavins by consuming NADH even if there is no styrene and/or the major epoxidase StyA1 present. Thus clearly shows this SMO can unproductively produce reduced flavin and so yielding uncoupling. The specific FAD reduction rate exceeds the observed epoxidation rates of StyA1 and StyA2-domain and allows serving sufficiently as the provider of reduced FAD of the system.

Sequence analysis and comparison of StyA1 and StyA2-fused with conventional SMO protein sequences revealed that all amino acid residues supposed to be involved in reduced FAD and substrate binding are conserved. And therefore, a similar reaction mechanism as described above for StyA from *P. putida* S12 was expected (Montersino et al. 2011). Very recent studies of StyA1 and StyA2B confirmed the drawn hypothesis (Tischler et al. 2011, 2013). Rapid mixing experiments of those proteins in oxygen saturated buffer with chemical reduced FAD and styrene resulted in the observation of similar absorption and fluorescence profiles as shown for StyA earlier (Kantz and Gassner 2011) indicating the reaction proceeds via the same FAD intermediates (Fig. 4.3) yielding finally styrene oxide, oxidized FAD, and H_2O. However, observed epoxidation rates of single-turnover experiments in dependence of styrene concentration (StyA1 1.6–52 s^{-1} and StyA2B 0.165–1.65 s^{-1}; Tischler et al. 2011) are much slower than those reported for StyA ($k = 104$ s^{-1}; Kantz and Gassner 2011) which is in congruence to the earlier reported steady-state data (Tischler et al. 2009, 2010). Another striking difference was observed. In case of StyA2B, the initial epoxidation occurred to be fast, but a highly fluorescent reaction intermediate built up during catalysis and was found to be the rate limiting step of StyA2B (Tischler et al. 2013). It is supposed to be the hydroxy-FAD intermediate formed as the first product after epoxidation (Fig. 4.3). It seemed to be stabilized by StyA2B, especially with excess styrene, and thus blocking the FAD-binding pocket for another round of epoxidation. This observation can explain the low epoxidation power of StyA2B observed earlier. But the structural reasons still need to be elucidated.

As mentioned above and shown in Fig. 4.1, the SMO type E2 comprises also members of the protein organization of StyA/StyB. However, for those enzymes only a single mechanistic study has been reported (Gröning et al. 2014). Herein, the reductase StyB origination from *Acinetobacter* sp. ADP1 had been studied in detail and a similar behavior as for StyA2B from strain 1CP was determined (NADH-dependent reduction of various flavins), but also some differences in the flavin binding had been measured. Furthermore, it is known for two representatives (ABQ12175, ABZ79366) found from metagenomes that they can convert indole to yield indigo as known for SMOs in general (Guan et al. 2007; Huang et al. 2008). Based on sequence similarity, they are close related to StyA1 of *R. opacus* 1CP and thus similar characteristics can be expected (Tischler et al. 2012).

It can be concluded the reaction mechanism of both SMO types are in general similar and proceed via the same intermediates. The nature and role of the natural fusion protein StyA2B still remains unclear, but first kinetic data allowed identifying the bottleneck in its oxidative half reaction. A major disadvantage of these enzymes is the unproductive production of reduced FAD and along that production of hydrogen peroxide (uncoupling). With respect to biotechnological applications of SMOs that issue has to be addressed and it might be solved by protein engineering or applying other (chemical) reductants in combination with design of experiments to optimize the process.

4.2 Styrene Oxide Isomerase

Styrene oxide isomerase (SOI, EC 5.3.99.7) represents the second enzyme of the upper pathway involved in side-chain oxidation of styrene and derivatives (Beltrametti et al. 1997; Bestetti et al. 2004; Itoh et al. 1997a; O'Connor et al. 1995; Panke et al. 1998). As an intramolecular oxidoreductase, the enzyme catalyzes the transformation of styrene oxide, which is formed by styrene monooxygenase, into phenylacetaldehyde. The formed product serves as a substrate for phenylacetaldehyde dehydrogenase. Styrene oxide isomerases were identified and partially investigated from several representatives of the genera *Xanthobacter*, *Pseudomonas*, *Rhodococcus*, *Corynebacterium*, and recently of *Sphingopyxis* and *Sphingobium* (Beltrametti et al. 1997; Hartmans et al. 1989; Itoh et al. 1997a; Oelschlägel et al. 2012a, 2014a, b, 2015b; Panke et al. 1998; Park et al. 2006;

Toda and Itoh 2012; Velasco et al. 1998). SOIs have commonly a native size of about 18 kDa and are encoded by *styC*-genes which are usually integrated in a *sty*-operon (see Chap. 3). Recently, a novel kind of SOI was identified from a *Sphingopyxis* strain with a size of 181 AA representing a 12–13 AA longer enzyme compared to SOIs from *Rhodococcus* and *Pseudomonas* strains and with a deviant substrate preference (Oelschlägel et al. 2014a, b, 2015a).

4.2.1 Mechanism of Substrate Isomerization

Racemic styrene oxide is converted into phenylacetaldehyde as the sole product without formation of by-products by SOIs investigated, so far (Oelschlägel et al. 2012a). That indicates a high selective transformation. Miyamoto et al. (2007) proposed a Meinwald rearrangement as potential mechanism (Fig. 4.4) (Meinwald et al. 1963). During the initial step, enzymatic protonation of the oxirane oxygen led to ring opening and the formation of a benzyl cation intermediate. The carbonium ion intermediate is thereby stabilized by a resonance effect of the π-electrons of the benzene ring. The intermediate undergoes enzymatic deprotonation into an

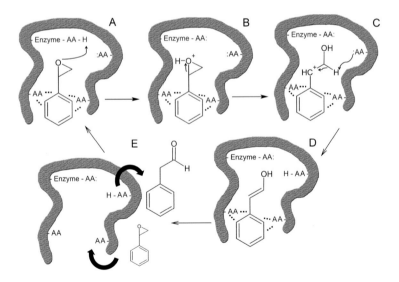

Fig. 4.4 Proposed mechanism of the styrene oxide isomerase (based on Miyamoto et al. 2007). Isomerization of chiral styrene oxide into the corresponding aldehyde is enantioselective with a preference of the (*S*)-enantiomer. Compared to (*R*)-styrene oxide, the (*S*)-stereoisomer is converted twice efficient (Itoh et al. 1997a; Miyamoto et al. 2007; Oelschlägel et al. 2012a). All SMOs described so far catalyzes a highly enantiospecific transformation of styrene into (*S*)-styrene oxide (Montersino et al. 2011) what makes the SOI-preference for (*S*)-styrene oxide reasonable. During isomerization, the chiral information gets lost in the case of conventional substrates like styrene oxide or ring-substituted derivatives

enol-form which is converted to an aldehyde by keto-enol tautomerization yielding the product and allows starting another catalytic cycle.

Substitution of styrene with electron-donating groups led to a faster transformation rate in the case of an SOI from *P. putida* S12 observed by Miyamoto et al. (2007). That fact as well as the absence of acetophenone as coproduct in the reaction catalyzed by another SOI of *R. opacus* 1CP strengthens that mechanism postulated (Oelschlägel et al. 2012a). Previous studies also disclosed a decrease of the relative activities by increasing van der Waals radii of different *p*-substituents, for example of fluorine, chlorine, bromine, as well as an influence of steric factors and hydrophilic binding effects (Oelschlägel et al. 2012a).

Recent investigations have also been predicting an important influence of some SOIs during the transformation of α-substituted styrene oxides (Oelschlägel et al. 2014b, 2015b). Transformation of 4-chloro-α-methylstyrene oxide, for example, with the biomass of *Pseudomonas fluorescens* ST, *R. opacus* 1CP, or a styrene-degrading *Sphingopyxis* allowed the production of the corresponding phenylacetic acid. In case of strain ST, enantiomeric excess of 80 % was reachable (Oelschlägel et al. 2014b) whereas strain 1CP and the *Sphingopyxis* strain Kp5.2 produced rather the racemic product or showed only a low preference of one enantiomer. That observation proposed an induced enantioselectivity by styrene monooxygenase which is maintained during transformation into the aldehyde by a probable site-specific attack of the SOI indicating the SOI has a stereoselective or even chiral active site. Because phenylacetaldehyde dehydrogenase oxidizes the aldehyde formed on a non-chiral carbon, it is proposed that the last step of the upper styrene degradation pathway does not affect the enantioselectivity of α-substituted substrates mentioned (Oelschlägel et al. 2014b).

4.2.2 SOI—A Membrane-Embedded Enzyme?

First, Hartmans et al. (1989) described difficulties during the enrichment of SOI from *Xanthobacter* sp. 124X by ion exchange or hydrophobic interaction chromatography indicating a distinct hydrophobic behavior of the enriched enzyme. Itoh et al. (1997a) suggested an integration of the SOI in the unsoluble cell-wall debris obtained during ultracentrifugation of crude extract from *Corynebacterium* strain AC-5. That was assumed because a treatment of the unsoluble SOI-containing fraction with several detergents, which are usually suitable for the solubilization of membrane proteins from the biological membrane, has been unsuccessful. Later investigations revealed that SOI is highly probable embedded in the cell membrane of the styrene-degrading bacteria *P. fluorescens* ST and *R. opacus* 1CP, respectively (Oelschlägel et al. 2012a). This study reported an efficient enrichment of the isomerase in the cell membrane debris after removal of cell-wall fragments and undisrupted whole cells by a centrifugation procedure. After removal of peripheral associated proteins by changing ionic strength in buffers, SOI had been enriched in the pellet containing the integral membrane proteins.

Fig. 4.5 A hypothetic membrane integration model of the styrene oxide isomerase from *R. opacus* 1CP is shown as cartoon. Similar transmembrane regions have been identified in other SOIs as well

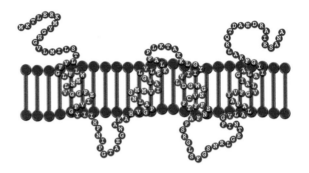

Treatment with detergents as CHAPS and guanidium chloride led to a high enrichment of the styrene oxide isomerase in the remained pellet, but solubilization of that enzyme was not achieved. SDS-PAGE-based investigations could prove changes in the behavior of the SOI after incubation with the detergent and the chaotrope salt guanidium chloride (Oelschlägel et al. 2012a). A solubilization of the SOI could be assumed under those conditions in a certain range, but an aggregation of the membrane-free hydrophobic SOI with itself or other similar proteins seems to lead to the formation of a non-soluble protein precipitate. Further, the assumption that SOI is an integral membrane-embedded enzyme was approved by the determination of hydrophobic transmembrane helices from available SOI protein sequences. Four strongly hydrophobic regions were determined in SOIs of *Pseudomonas*, *Rhodococcus*, and *Sphingopyxis* strains (Fig. 4.5) (Oelschlägel et al. 2014a, b) and explain the difficulties and the behavior of those enzymes during enrichment. Latest studies also show potential enrichment of SOI in the membrane fraction for representatives of the genera *Sphingopyxis* and *Sphingobium* (Oelschlägel et al. 2014a, b). But, the SOI of *Xanthobacter* sp. strain 124X seems to show an untypical behavior during purification compared to other SOIs mentioned and a cell-wall-embedded localization of that enzyme could not ruled out. Additionally, larger differences in the gene sequence of that SOI compared to other ones were assumed by PCR-approaches to determine the respective sequence (Oelschlägel et al. 2014a, b). These findings reinforce earlier made assumptions on the special behavior and thus position among related enzymes of this SOI originating from strain 124X (Hartmans et al. 1989).

Membrane localization clearly stabilizes SOI toward several environmental conditions. An enriched enzyme preparation of *R. opacus* 1CP tolerates high temperatures of up to 50 °C without significant inactivation and a maximum activity was determined at 65 °C (Oelschlägel et al. 2012a). For another SOI of *Corynebacterium* sp. AC-5, a similar behavior was reported with an optimum of 40 °C (Itoh et al. 1997a). Remarkably, some SOIs could be activated by single heat treatment. After an incubation of various enzymes which had been enriched from *R. opacus* 1CP and *P. fluorescens* ST at 40 °C for 30 min, SOI activity increased up to 222 % of the initial enzyme activity (Oelschlägel et al. 2012a). Highest pH stability and activity optima had been determined to be at pH around 7 for the enzyme of *R. opacus* 1CP and *Corynebacterium* sp. AC-5.

A recent study reported a high stability of the isomerase of strain 1CP toward detergents and chaotropic agents. The enzyme tolerates a 30 min-incubation in presence of up to 0.3 % (w/v) SDS, 2 % (w/v) CHAPS, or up to 3.5 M guanidinium chloride without significant loss of activity (Oelschlägel et al. 2012a). Also high storability was demonstrated at −20 °C without loss of activity over months and at 2–3 °C with a decrease to 70 % of initial activity during 10 weeks. A high storability was also shown by Hartmans et al. (1989). The proposed membrane localization is important for the stability of those enzymes and had also been proved by a deviant stability after treatment of SOI with CHAPS and guanidinium chloride. After incubation under conditions mentioned, enzyme preparation was less stable and was inactivated after some weeks at −20 °C (Oelschlägel et al. 2012a). Despite a high stability of the enzyme toward some chemicals and temperature, the product phenylacetaldehyde irreversibly inhibits the enzyme at higher concentrations (Oelschlägel et al. 2012a). In the presence of 55 mM phenylacetaldehyde, the half-life of this SOI was determined to be about 15 min, which is for a biotechnological process a limiting factor of course.

Membrane localization also offers the advantage to easily enrich the enzyme from crude extract by simple centrifugation and treatment with buffers of deviating ion strength which was patented (Oelschlägel et al. 2012b). High SOI activities in the crude or cell-free extract were reported for *P. fluorescens* ST, *R. opacus* 1CP, *Xanthobacter* sp. 124X, and *Corynebacterium* sp. AC-5 with 7.6–20.7 U mg^{-1} after induction with suitable inducers representing substrates and intermediates of side-chain oxidation (Itoh et al. 1997a; Oelschlägel et al. 2012a, 2014a, b). Specific activities in the crude extract could be enriched by the factor 28–42 up to 313–370 U mg^{-1} as described by Oelschlägel et al. (2012a).

4.3 Phenylacetaldehyde Dehydrogenase

Phenylacetaldehyde dehydrogenase (EC 1.2.1.39; PAD, FeaB) is the third enzyme of the styrene specific side-chain oxygenation route (see Chaps. 2 and 3), but, can also be part of other metabolic pathways as exemplary shown for the degradation of ethylbenzene, phenylalanine, 2-phenylethylamine, and 2-phenylethanol (Arias et al. 2008; Corkery et al. 1994; Ferrández et al. 1997; Hanlon et al. 1997; Hartmans et al. 1989; Parrott et al. 1987; Rodríguez-Zavala et al. 2006; Shimizu et al. 1993). In all cases it yields phenylacetic acid as the central metabolite. However, there are only limited mechanistic and structural studies on these enzymes, rather there are reports on general activity available (Beltrametti et al. 1997; Cox et al. 1996; Hartmans et al. 1989, 1990; O'Connor et al. 1995; Oelschlägel et al. 2012a), whereas, often it has been shown that different microorganisms utilize styrene and yield phenylacetic acid as metabolite thus it was reasoned that PAD is active in these microorganisms (Oelschlägel et al. 2014a, 2015a; Panke et al. 1998; Park et al. 2006; Tischler et al. 2009; Toda and Itoh 2012; Velasco et al. 1998). Nikodinovic-Runic et al. (2009) investigated the mentioned styrene pathway and investigated the transcription of respective genes and thus protein production. Results clearly indicate that SMOA and PAD are the most abundant proteins of the pathway under styrene inducing conditions. The high level of PAD might be explained simply since the substrate, phenylacetaldehyde, needs to be detoxified (Ferrández et al. 1997; Parrott et al. 1987). Otherwise, it could accumulate in the cytoplasm and inactivate enzymes or alter nucleotides by making covalent linkages. Changing from the non-limiting to nitrogen-limiting conditions did not affect PAD levels in the cell (Nikodinovic-Runic et al. 2009). Furthermore, in case of *R. opacus* 1CP, a constitutive expression of PAD-gene was supposed, since the PAD activity determined was independent of the carbon source supplied during the cultivation of the respective strain (Oelschlägel et al. 2012a).

The phenylacetaldehyde dehydrogenase catalyzes the nonreversible reaction of phenylacetaldehyde and other aromatic aldehydes into the corresponding acids by the expense of NAD^+ and H_2O (Cox et al. 1996; Ferrández et al. 1997; Rodríguez-Zavala et al. 2006). NADH is formed as by-product and released into the medium and is being supposed to be used rapidly by other enzymes. Taking the complete activity of the upper styrene degradation pathway into account as an enzymatic cascade (Chap. 2) it might be reasoned that SMO and PAD act in concert to avoid wasting reducing equivalents. Thus the SMO utilizes NADH and provides NAD^+ which is recycled to NADH by the PAD. SOI does not utilize cofactors and is therefore not included. However, PAD is not able to act on 2-phenylethanol indicating a direct oxidation of phenylacetaldehyde into phenylacetic acid (Beltrametti et al. 1997). For some enzymes, it has been shown that they accept besides NAD^+ also $NADP^+$ and/or phenazine methosulfate (PMS) as the final electron acceptor (Corkery et al. 1994; Ferrández et al. 1997; Hartmans et al. 1989, 1990; Rodríguez-Zavala et al. 2006). Interestingly, these dehydrogenases are highly specific for phenylacetaldehyde and only related compounds as halogenated derivatives are converted. Aldehyde dehydrogenases typically have an esterase activity, but PAD does not possess that activity in a remarkable manner since it converted *p*-nitrophenol acetate only at a rate <1 % compared to the dehydrogenase activity (Ferrández et al. 1997; Rodríguez-Zavala et al. 2006). The detailed mechanism of the enzyme remains still unclear and amino acid residues which are necessary to bind substrate and define specificity are not well characterized. Thus mechanistic and structural studies are needed to uncover the biochemistry of PAD.

4.4 Phenylacetaldehyde Reductase

2-Phenyletahnol was determined as a key-metabolite during the styrene degradation by means of several bacteria (see Sect. 2.2.4) (Hartmans et al. 1989; Marconi et al. 1996; Utkin et al. 1991). It might be obtained directly from styrene oxide or from phenylacetaldehyde. The latter seems most favorable (Tischler and Kaschabek 2012), and thus we provide an overview on the respective enzyme phenylacetaldehyde reductase (PAR). Respectively, the systematic name according EC nomenclature is

2-phenylethanol: NAD^+ oxidoreductase, but PAR is more often used in the literature. Such an enzymatic activity has in addition been found in fungi (Gopalakrishna et al. 1976) and plants (Chen et al. 2011; Tieman et al. 2007). However, in those cases it seems more a productive (anabolic) than a degradative (catabolic) route.

In case of *Rhodococcus* sp. ST-10 (previously designated as *Corynebacterium* sp. ST-10) PAR was supposed to be a key enzyme of the styrene degradation pathway (Itoh et al. 1997b) and for the first time PAR has been investigated from the wild-type organism (Itoh et al. 1997b) as well as in a recombinant form (Wang et al. 1999a, b). Other isolates (strains ST-4, ST-5, and AC-5) known to utilize styrene also showed this enzymatic activity. In *Rhodococcus* sp. ST-10 it was induced and produced up on the presence of styrene or styrene oxide, and after purification it could be shown that PAR accepts many substrates besides phenylacetaldehyde. Interestingly, it does not catalyze the reverse reaction to produce phenylacetaldehyde from 2-phenylethanol which is probably the reason for naming it PAR instead of 2-phenylethanol: NAD^+ oxidoreductase. This finding might be caused by an observed product inhibition and thus not enough phenylethanol can be provided to overcome kinetic barriers. However, later it has been demonstrated that PAR is able to convert small alcohols into corresponding ketones (Itoh et al. 2002). Thus it was demonstrated PAR is able to convert aldehydes and alcohols and therefore a true alcohol oxidoreductase. Further, it solely accepts NADH as a cofactor which is in congruence to the SMO (StyB) of the same pathway (Itoh et al. 1997b). The active site PAR has an NAD^+- and two zinc-binding sites participating in the catalytic cycle (Wang et al. 1999a, b) and it was determined as homotetramer with almost 8 mol zinc per mol tetramer which means each subunit is loaded with two zinc ions, respectively. As many dehydrogenases PAR is a chiral agent. It converts acetophenone into *S*-1-phenylethanol with more than 96 % ee and in addition allows the chiral conversion of many other substrates (Itoh et al. 1997b, 1999). Therefore, it was presumed that PAR in combination with NADH spans an oriented catalytic site allowing chiral turnover of those substrates making PAR an interesting biocatalyst.

No structural data are available for PAR, but, for a related dehydrogenase from *Rhodococcus ruber* DSM 44541 (PDB: 2XAA; Karabec et al. 2010). The general protein organization seems highly similar when comparing the amino acid sequences of both PAR and 2XAA. However, due to many mutagenetic studies further information on key residues of PAR are available (Makino et al. 2005, 2007; Wang et al. 1999a). Further biochemical studies are necessary to understand the mechanism and especially to enlighten the product inhibition not allowing the production of phenylacetaldehyde by the reverse reaction starting with 2-phenylethanol.

4.5 Epoxide Hydrolase

During styrene degradation by fungi (Braun-Lüllemann et al. 1997) as well as in human styrene detoxification (Rueff et al. 2009), a similar route is followed: styrene—styrene oxide—styreneglycol—mandelic acid. Mandelic acid likely gets oxidatively decarboxylated to benzoic acid which is a central intermediate and can be funneled into the metabolism, respectively. A major difference to the bacterial side-chain oxygenation pathway is the activity of a styrene oxide hydrolase (epoxide hydrolase) (see Sect. 2.2.4). However, not much about those degradative enzymes is known (Braun-Lüllemann et al. 1997; Rustemov et al. 1992) in comparison to the enzymes involved in the mammalian detoxification processes (Mosisseau and Hammock 2005; Orru and Faber 1999; Tischler and Kaschabek 2012).

The expoxide hydrolases (EC 3.3.2.10; EH) do in general catalyze a two-step reaction from an oxirane to a vicinal diol and are independent of any cofactor (Mosisseau and Hammock 2005; Kotik et al. 2012; Orru and Faber 1999). Initially, the oxirane gets attacked by a nucleophile (e.g., aspartate) to yield a covalent intermediate designated "glycol-monoester intermediate." In order to perform the ester hydrolysis, H_2O gets activated by a histidine which abstracts a proton to generate a stronger nucleophile: HO^-. The latter attacks the ester intermediate and allows the formation of the final vicinal diol while having the enzyme in its starting configuration for another catalytic cycle. Respectively, in dependence on the carbon atom of the oxirane which gets attacked first, the stereoselectivity of the reaction can differ. Thus a retention and/or inversion are possible. The configuration obtained upon oxirane conversion is strongly dependent on the enzyme and its active site as well as on the substitutional pattern of the substrate to be converted. However, in case of styrene oxide-like substrates, a special property needs to be mentioned. The formal presence of a benzylic carbon at the styrene oxide supports the formation of a carbo cation which is stabilized due to the aromatic system. In case of styrene oxide, both carbon atoms of the oxirane system can easily be attacked since one is electronically and the other one sterically more attractive (Orru and Faber 1999).

Racemic styrene oxide is a model compound in order to evaluate the enantioselective hydrolysis of oxiranes by epoxide hydrolases (Liu et al. 2006; Kim et al. 2006; Kotik et al. 2012; Orru and Faber 1999). Only one of the enantiomers of the racemic substrate serves as a substrate. So it is possible to produce enantioenriched epoxides and vicinal diols. Often, the S-enantiomer is formed with a high enantioselectivity >98 % (Lee et al. 2004; Liu et al. 2006; Kim et al. 2006).

References

Arias S, Olivera ER, Arcos M, Naharro G, Luengo JM (2008) Genetic analyses and molecular characterization of the pathways involved in the conversion of 2-phenylethylamine and 2-phenylethanol into phenylacetic acid in *Pseudomonas putida* U. Environ Microbiol 10:413–432

Beltrametti F, Marconi AM, Bestetti G, Galli E, Ruzzi M, Zennaro E (1997) Sequencing and functional analysis of styrene catabolism genes from *Pseudomonas fluorescens* ST. Appl Environ Microbiol 63:2232–2239

Bestetti G, Di Gennaro P, Colmenga A, Ronco I, Galli E, Sello G (2004) Characterization of styrene catabolic pathway in *Pseudomonas fluorescens* ST. Int. Biodeterior Biodegradation 54:183–187

Braun-Lüllemann A, Majcherczyk A, Huttermann A (1997) Degradation of styrene by white-rot fungi. Appl Microbiol Biotechnol 47:150–155

Chen X-M, Kobayashi H, Sakai M, Hirata H, Asai T, Ohnishi T, Baldermann S, Watanabe N (2011) Functional characterization of rose phenylacetaldehyde reductase (PAR), an enzyme involved in the biosynthesis of the scent compound 2-phenylethanol. J Plant Physiol 168:88–95

Corkery AM, O'Connor KE, Buckley CM, Dobson ADW (1994) Ethylbenzene degradation by Pseudomonas fluorescens strain CA-4. FEMS Microbiol Lett 124:23–28

Cox HHJ, Faber BW, van Heiningen WNM, Radhoe H, Doddema HJ, Harder W (1996) Styrene metabolism in Exophiala jeanselmei and involvement of a cytochrome P-450-dependent styrene monooxygenase. Appl Environ Microbiol 62:1471–1474

Feenstra KA, Hofstetter K, Bosch R, Schmid A, Commandeur JNM, Vermeulen NPE (2006) Enantioselective substrate binding in a monooxygenase protein model by molecular dynamics and docking. Biophys J 91:3206–3216

Ferrández A, Prieto MA, García JL, Díaz E (1997) Molecular characterization of PadA, a phenylacetaldehyde dehydrogenase from Escherichia coli. FEBS Lett 406:23–27

Gopalakrishna Y, Narayanan TK, Ramanada Rao G (1976) Biosynthesis of β-phenethyl alcohol in Candida guilliermondii. Biochem Biophys Res Commun 69:417–422

Gröning JAD, Kaschabek SR, Schlömann M, Tischler D (2014) A mechanistic study on SMOB-ADP1: an NADH:flavin oxidoreductase of the two-component styrene monooxygenase of Acinetobacter baylyi ADP1. Arch Microbiol 196:829–845

Guan C, Ju J, Borlee BR, Williamson LL, Shen B, Raffa KF, Handelsman J (2007) Signal mimics derived from a metagenomic analysis of the gypsy moth gut microbiota. Appl Environ Microbiol 73:3669–3676

Gursky LJ, Nikodinovic-Runic J, Feenstra KA, O'Connor KE (2010) In vitro evolution of styrene monooxygenase from Pseudomonas putida CA-3 for improved epoxide synthesis. Appl Microbiol Biotechnol 85:995–1004

Hanlon SP, Hill TK, Flavell MA, Stringfellow JM, Cooper RA (1997) 2-Phenylethylamine catabolism by Escherichia coli K-12: gene organization and expression. Microbiology 143:513–518

Hartmans S, Smits JP, van der Werf MJ, Volkering F, de Bont JAM (1989) Metabolism of styrene oxide and 2-phenylethanol in the styrene-degrading Xanthobacter strain 124X. Appl Environ Microbiol 55:2850–2855

Hartmans S, van der Werf MJ, De Bont JAM (1990) Bacterial degradation of styrene involving a novel flavin adenine dinucleotide-dependent styrene monooxygenase. Appl Environ Microbiol 56:1347–1351

Hollmann F, Lin P-C, Witholt B, Schmid A (2003) Stereospecific biocatalytic epoxidation: the first example of direct regeneration of a FAD-dependent monooxygenase for catalysis. J Am Chem Soc 125:8209–8217

Huang YL, Zhang J, Zhou SN (2008) Identification and production of indigo retrieved from a deep-sea sediment metagenomic library. Submitted to GenBank: ABZ79366. Article unpublished

Huijbers MME, Montersino S, Westphal AH, Tischler D, van Berkel WJH (2014) Flavin dependent monooxygenases. Arch Biochem Biophys 544:2–17

Itoh N, Hayashi K, Okada K, Ito T, Mizuguchi N (1997a) Characterization of styrene oxide isomerase, a key enzyme of styrene and styrene oxide metabolism in Corynebacterium sp. Biosci Biotechnol Biochem 61:2058–2062

Itoh N, Morihama R, Wang J, Okada K, Mizuguchi N (1997b) Purification and characterization of phenylacetaldehyde reductase from a styrene-assimilating Corynebacterium strain, ST-10. Appl Environ Microbiol 63:3783–3788

Itoh N, Mizuguchi N, Mabuchi M (1999) Production of chiral alcohols by enantioselective reduction with NADH-dependent phenylacetaldehyde reductase from Corynebacterium strain, ST-10. J Mol Catal B Enzym 6:41–50

Itoh N, Matsuda M, Mabuchi M, Dairi T, Wang J (2002) Chiral alcohol production by NADH-dependent phenylacetaldehyde reductase coupled with in situ regeneration of NADH. Eur J Biochem 269:2394–2402

Kantz A, Gassner GT (2011) Nature of the reaction intermediates in the flavin adenine dinu-cleotide-dependent epoxidation mechanism of styrene monooxygenase. Biochemistry 50:523–532

Kantz A, Chin F, Nallamothu N, Nguyen T, Gassner GT (2005) Mechanism of flavin transfer and oxygen activation by the two-component flavoenzyme styrene monooxygenase. Arch Biochem Biophys 442:102–116

Karabec M, Łyskowski A, Tauber KC, Steinkellner G, Kroutil W, Grogan G, Gruber K (2010) Structural insights into substrate specificity and solvent tolerance in alcohol dehydrogenase ADH-'A' from *Rhodococcus ruber* DSM 44541. Chem Commun 46:6314–6316

Kim HS, Lee OK, Lee SJ, Hwang S, Kim SJ, Yang SH, Park S, Lee EY (2006) Enantioselective epoxide hydrolase activity of a newly isolated microorganism, *Sphingomonas echinoides* EH-983, from seawater. J Mol Catal B Enzym 41:130–135

Kotik M, Archelas A, Wohlgemuth R (2012) Epoxide hydrolases and their application in organic synthesis. Curr Org Chem 16:451–482

Lee EY, Yoo S-S, Kim HS, Lee SJ, Oh Y-K, Park S (2004) Production of (*S*)-styrene oxide by recombinant *Pichia pastoris* containing epoxide hydrolase from *Rhodotorula glutinis*. Enzyme Microbiol Technol 35:624–631

Lin H, Qiao J, Liu Y, Wu Z-L (2010) Styrene monooxygenase from *Pseudomonas* sp. LQ26 cata-lyzes the asymmetric epoxidation of both conjugated and unconjugated alkenes. J Mol Catal B Enzym 67:236–241

Lin H, Tang D-F, Qaed Ahmed AA, Liu Y, Wu Z-L (2012) Mutations at the putative active cavity of styrene monooxygenase: enhanced activity and reversed enantioselectivity. J Biotechnol 161:235–241

Liu Z, Michel J, Wang Z, Witholt B, Li Z (2006) Enantioselective hydrolysis of styrene oxide with the epoxide hydrolase of *Sphingomonas* sp. HXN-200. Tetra-hedron: Asymmetry 17:47–52

Makino Y, Inoue K, Dairi T, Itoh N (2005) Engineering of phenylacetaldehyde reductase for efficient substrate conversion in concentrated 2-propanol. Appl Environ Microbiol 71:4713–4720

Makino Y, Dairi T, Itoh N (2007) Engineering the phenylacetaldehyde reductase mutant for improved substrate conversion in the presence of concentrated 2-propanol. Appl Microbiol Biotechnol 77:833–843

Marconi AM, Beltrametti F, Bestetti G, Solinas F, Ruzzi M, Galli E, Zennaro E (1996) Cloning and characterization of styrene catabolism genes from *Pseudomonas fluorescens* ST. Appl Environ Microbiol 62:121–127

Massey V (1994) Activation of molecular oxygen by flavins and flavoproteins. J Biol Chem 269:22459–22462

Meinwald J, Labana SS, Chadha MSJ (1963) Peracid reactions. III. The oxidation of bicyclo [2.2.1] heptadiene. J Am Chem Soc 85:582–585

Miyamoto K, Okuro K, Ohta H (2007) Substrate specificity and reaction mechanism of recombi-nant styrene oxide isomerase from *Pseudomonas putida* S12. Tetrahedron Lett 48:3255–3257

Montersino S, Tischler D, Gassner GT, van Berkel WJH (2011) Catalytic and structural features of flavoprotein hydroxylases and epoxidases. Adv Synth Catal 353:2301–2319

Mooney A, Ward PG, O'Connor KE (2006) Microbial degradation of styrene: biochemistry, molecular genetics, and perspectives for biotechnological applications. Appl Microbiol Biotechnol 72:1–10

Morisseau C, Hammock BD (2005) Epoxide hydrolases: mechanisms, inhibitor designs, and bio-logical Roles. Annu Rev Pharmacol Toxicol 45:311–333

Morrison E, Kantz A, Gassner GT, Sazinsky MH (2013) Structure and mechanism of styrene monooxygenase reductase: new insight into the FAD-transfer reaction. Biochemistry 52:6063–6075

Nikodinovic-Runic J, Flanagan M, Hume AR, Cagney G, O'Connor KE (2009) Analysis of the *Pseudomonas putida* CA-3 proteome during growth on styrene under nitrogen-limiting and non-limiting conditions. Microbiology 155:3348–3361

O'Connor K, Buckley CM, Hartmans S, Dobson AD (1995) Possible regulatory role for non-aromatic carbon sources in styrene degradation by *Pseudomonas putida* CA-3. Appl Environ Microbiol 61:544–548

O'Connor KE, Dobson AD, Hartmans S (1997) Indigo formation by microorganisms expressing styrene monooxygenase activity. Appl Environ Microbiol 63:4287–4291

Oelschlägel M, Gröning JAD, Tischler D, Kaschabek SR, Schlömann M (2012a) Styrene oxide isomerase of *Rhodococcus opacus* 1CP, a highly stable and considerably active enzyme. Appl Environ Microbiol 78:4330–4337

Oelschlägel M, Tischler D, Gröning JAD, Kaschabek SR, Schlömann M (2012b) Process for the enzymatic synthesis of aromatic aldehydes or ketones. Patent: DE 102011006459 A1 20121004

Oelschlägel M, Zimmerling J, Schlömann M, Tischler D (2014a) Styrene oxide isomerase of *Sphingopyxis* sp. Kp5.2. Microbiol (UK) 160:2481–2491

Oelschlägel M, Zimmerling J, Tischler D, Schlömann M (2014b) Method for biocatalytic synthesis of substituted or unsubstituted phenylacetic acids and ke-tones having enzymes of microbial styrene degradation. Patent: DE 102013211075 A1 20141218; WO 2014198871 A2 20141218

Oelschlägel M, Rückert C, Kalinowski J, Schmidt G, Schlömann M, Tischler D (2015a) Description of *Sphingopyxis fribergensis* sp. nov.—a soil bacterium with the ability to degrade styrene and phenylacetic acid. Int J Syst Evol Microbiol. doi:10.1099/ijs.0.000371

Oelschlägel M, Kaschabek SR, Zimmerling J, Schlömann M, Tischler D (2015b) Co-metabolic formation of substituted phenylacetic acids by styrene-degrading bacteria. Biotechnol Rep 6:20–26

O'Leary ND, O'Connor KE, Dobson ADW (2002) Biochemistry, genetics and physiology of microbial styrene degradation. FEMS Microbiol Rev 26:403–417

Orru RV, Faber K (1999) Stereoselectivities of microbial epoxide hydrolases. Curr Opin Chem Biol 3:16–21

Otto K, Hofstetter K, Roethlisberger M, Witholt B, Schmid A (2004) Biochemical characterization of StyAB from *Pseudomonas* sp. strain VLB120 as a two-component flavin-diffusible monooxygenase. J Bacteriol 186:5292–5302

Panke S, Witholt B, Schmid A, Wubbolts MG (1998) Towards a biocatalyst for (*S*)-styrene oxide production: characterization of the styrene degradation pathway of *Pseudomonas* sp. strain VLB120. Appl Environ Microbiol 64:2032–2043

Park MS, Bae JW, Han JH, Lee EY, Lee S-G, Park S (2006) Characterization of styrene catabolic genes of *Pseudomonas putida* SN1 and construction of a recombinant Escherichia coli containing styrene monooxygenase gene for the production of (*S*)-styrene oxide. J Microbiol Biotechnol 16:1032–1040

Parrott S, Jones S, Cooper RA (1987) 2-Phenylethylamine catabolism by *Escherichia coli* K12. J Gen Microbiol 133:347–351

Paul CE, Tischler D, Riedel A, Heine T, Itoh N, Hollmann F (2015) Nonenzymatic regeneration of styrene monooxygenase for catalysis. ACS Catal 5:2961–2965

Qaed AA, Lin H, Tang DF, Wu Z-L (2011) Rational design of styrene monooxygenase mutants with altered substrate preference. Biotechnol Lett 33:611–616

Riedel A, Heine T, Westphal AH, Conrad C, van Berkel WJH, Tischler D (2015) Catalytic and hydrodynamic properties of styrene monooxygenases from *Rhodococcus opacus* 1CP are modulated by cofactor binding. AMB Express 5:30

Rodríguez-Zavala JS, Allali-Hassani A, Weiner H (2006) Characterization of *E. coli* tetrameric aldehyde dehydrogenases with atypical properties compared to other aldehyde dehydrogenases. Protein Sci 15:1387–1396

Rueff J, Teixeira JP, Santos LS, Gaspar JF (2009) Genetic effects and biotoxi-city monitoring of occupational styrene exposure. Clin Chim Acta 399:8–23

Rustemov SA, Golovleva LA, Alieva RM, Baskunov BP (1992) New pathway of styrene oxidation by a *Pseudomonas putida* culture. Microbiologica 61:1–5

Shimizu E, Ichise H, Odawara T, Yorifuji T (1993) NADP-dependent phenylacetaldehyde dehydrogenase for degradation of phenylethylamine in *Arthrobacter globiformis*. Biosci Biotech Biochem 57:852–853

Tieman DM, Loucas HM, Kim JY, Clark DG, Klee HJ (2007) Tomato phenylacetaldehyde reductases catalyze the last step in the synthesis of the aroma volatile 2-phenylethanol. Phytochem 68:2660–2669

Tischler D, Kaschabek SR (2012) Microbial degradation of xenobiotics. SN Singh (ed). Springer, Berlin, p 67–99

Tischler D, Eulberg D, Lakner S, Kaschabek SR, van Berkel WJH, Schlömann M (2009) Identification of a novel self-sufficient styrene monooxygenase from *Rhodococcus opacus* 1CP. J Bacteriol 191:4996–5009

Tischler D, Kermer R, Gröning JAD, Kaschabek SR, van Berkel WJH, Schlömann M (2010) StyA1 and StyA2B from *Rhodococcus opacus* 1CP: A multifunctional styrene monooxygenase system. J Bacteriol 192:5220–5227

Tischler D, Kaschabek SR, Gassner GT (2011) StyA1/StyA2B, a unique flavin monooxygenase system. In: Proceedings of the seventeenth international symposium on flavins and flavoproteins, Berkeley, p 307–312

Tischler D, Gröning JAD, Kaschabek SR, Schlömann M (2012) One-component styrene monooxygenases: an evolutionary view on a rare class of flavoproteins. Appl Biochem Biotechnol 167:931–944

Tischler D, Schlömann M, van Berkel WJH, Gassner GT (2013) FAD C(4a)-hydroxide stabilized in a naturally fused styrene monooxygenase. FEBS Lett 587:3848–3852

Toda H, Itoh N (2012) Isolation and characterization of styrene metabolism genes from styrene-assimilating soil bacteria *Rhodococcus* sp. ST-5 and ST-10. J Biosci Bioeng 113:12–19

Toda H, Imae R, Komio T, Itoh N (2012) Expression and characterization of styrene monooxygenases of *Rhodococcus* sp. ST-5 and ST-10 for synthesizing enantiopure (*S*)-epoxides. Appl Microbiol Biotechnol 96:407–418

Ukaegbu UE, Kantz A, Beaton M, Gassner GT, Rosenzweig AC (2010) Structure and ligand binding properties of the epoxidase component of styrene monooxygenase. Biochemistry 49:1678–1688

Utkin I, Yakimov M, Matveeva L, Kozlyak E, Rogozhin I, Solomon Z, Bez-borodov A (1991) Degradation of styrene and ethylbenzene by *Pseudomonas* species Y2. FEMS Microbiol Lett 77:237–242

van Berkel WJH, Kamerbeek NM, Fraaije MW (2006) Flavoprotein monooxygenases, a diverse class of oxidative biocatalysts. J Biotechnol 124:670–689

van den Heuvel RHH, Westphal AH, Heck AJR, Walsh MA, Rovida S, van Berkel WJH, Mattevi A (2004) Structural studies on flavin reductase PheA2 reveal binding of NAD in an unusual folded conformation and support novel mechanism of action. J Biol Chem 279:12860–12867

van Hellemond EW, Janssen DB, Fraaije MW (2007) Discovery of a novel styrene monooxygenase originating from the metagenome. Appl Environ Microbiol 73:5832–5839

Velasco A, Alonso S, Garcia JL, Perera J, Diaz E (1998) Genetic and functional analysis of the styrene catabolic cluster of *Pseudomonas* sp. strain Y2. J Bacteriol 180:1063–1071

Volmer J, Neumann C, Bühler B, Schmid A (2014) Engineering of *Pseudomonas taiwanensis* VLB120 for constitutive solvent tolerance and increased specific styrene epoxidation activity. Appl Environ Microbiol 80:6539–6548

Wang J-C, Sakakibara M, Matsuda M, Itoh N (1999a) Site-directed mutagenesis of two zinc-binding centers of the NADH-dependent phenylacetaldehyde reductase from styrene-assimilating *Corynebacterium* sp. strain ST-10. Biosci Biotechnol Biochem 63:2216–2218

Wang J-C, Sakakibara M, Liu J-Q, Dairi T, Itoh N (1999b) Cloning, sequence analysis, and expression in *Escherichia coli* of the gene encoding phenylacetaldehyde reductase from styrene-assimilating *Corynebacterium* sp. strain ST-10. Appl Microbiol Biotechnol 52:386–392

Chapter 5
Biotechnological Applications of Styrene-Degrading Microorganisms or Involved Enzymes

Abstract The development of sustainable technologies which are more ecofriendly as common alternatives is a major goal of biotechnology. In this context, microorganisms and their repertoire of biocatalysts serve as a resource in many respects. Styrene-degrading microorganisms and their enzymes also can serve as such a rich resource for biotechnology. The microorganisms, for example, tolerate organic solvents, degrade recalcitrant and toxic compounds, enrich valuable intermediates, and can be genetically manipulated. Thus these can be employed to treat waste streams in order to detoxify polluted air and aquifer. Their regulatory network can also be applied as a form of a biosensor to report on the presence of such toxic compounds. Or the biomass can be used as whole-cell biocatalyst in order to produce valuable compounds such as styrene derivatives for polymer synthesis, indigoid dyes, aroma compounds, or even pharmaceuticals as ibuprofen. The reservoir of biocatalysts can also be used as a source for genetic material. Thus genes can be manipulated and cloned to (recombinantly) produce heterologous enzymes for the biocatalysis in alternative hosts as *Escherichia coli*. Often these enzymes are later employed to produce (enantiopure) compounds which serve as valuable building blocks for various fields as exemplary aroma production, agrochemistry, food and feed, pharmaceutical production, or polymer chemistry.

Keywords Styrene monooxygenase · Styrene oxide isomerase · Biosensor · Bioremediation · Enzyme cascade · Styrene polymer · Phenylethanol

© The Author(s) 2015
D. Tischler, *Microbial Styrene Degradation*,
SpringerBriefs in Microbiology, DOI 10.1007/978-3-319-24862-2_5

5.1 Styrene Monooxygenase, a Versatile and Powerful Biocatalyst

The ability of styrene monooxygenases (SMOs) to perform highly regio- and enantioselective monooxidations of alkenes and sulfides is a feature focussing these enzymes into chemical industry as reviewed by Huijbers et al. (2014), Lechner et al. (2015), Montersino et al. (2011), Schmid et al. (2001), and van Berkel et al. (2006). A large number of possible substrates allow efficient production of valuable building blocks or compounds for fine chemical and pharmaceutical industries, agrochemical uses, or even for dye industry (Table 5.1).

In the recent years, two major directions had been investigated: (a) whole-cell and (b) cell-free biocatalysis. When whole cells expressing SMOs have to

Table 5.1 Products obtained from SMO-based catalysis

	Product(s)	Origin	Reference(s)
SMO-driven approaches (in few cases including a regeneration system for NADH/FAD)			
StyA/B, StyA/chem., StyA1/A2B	Styrene oxide and derivatives	Various strains of *Pseudomonas* spp. and *Rhodococcus* spp., metagenome	Bernasconi et al. (2000, 2004), Gursky et al. (2010), Hollman et al. (2003), Lin et al. (2010), Otto et al. (2004), Panke et al. (2002), Qaed et al. (2011), Tischler et al. (2009, 2010), Toda et al. (2012a, b), van Hellemond et al. (2007)
StyA/B, StyA/chem., StyA1/A2B	Cyclic epoxides	Various strains of *Pseudomonas* spp. and *Rhodococcus* spp.	Bernasconi et al. (2000, 2004), Gursky et al. (2010), Hollman et al. (2003), Lin et al. (2010, 2012), Tischler et al. (2010), Toda et al. (2012a, b)
StyA/B	Aliphatic epoxides	*Rhodococcus* sp. ST-10, *Rhodococcus* sp. ST-5	Toda et al. (2012a, b, 2014, 2015)
StyAB2	Glycidol derivatives	*Pseudomonas* sp. LQ26	Lin et al. (2011)
StyA/B; StyAB2	Other epoxides	*Pseudomonas* sp. LQ26, *Pseudomonas fluorescens* ST, *Rhodococcus* sp. ST-10, *Rhodococcus* sp. ST-5	Bernasconi et al. (2000), Di Gennaro et al. (1999), Lin et al. (2010, 2011), Toda et al. (2012a, b)

(continued)

Table 5.1 (continued)

	Product(s)	Origin	Reference(s)
StyA/B, StyA1/StyB, StyA1/chem., SMO	Aromatic sulfoxides	*Pseudomonas tai-wanensis VLB120, P. putida* CA-3, *R. opacus* 1CP, metagenome	Boyd et al. (2012), Hollman et al. (2003), Nikodinovic-Runic et al. (2013), Paul et al. (2015), Riedel et al. (2015), Tischler et al. (2010), Toda et al. (2012a), van Hellemond et al. (2007)
Combinatorial approaches (often the host cell adds an activity)			
Host cell, SMO, SOI	Indigo	*P. putida* CA-3, *R. opacus* 1CP, *Rhodococcus* sp. ST-10, *Rhodococcus* sp. ST-5, metagenome	Guan et al. (2007), Gursky et al. (2010), O'Connor et al. (1997), Tischler et al. (2009), Toda et al. (2012b), van Hellemond et al. (2007)
StyA, styrene dioxygenase	(*R*)-1,2-phenyl-ethanediol	Recombinant	McKenna et al. (2013)
StyA, NDDH	Epoxy cinnamic acid	Recombinant	Di Gennaro et al. (2013)
sty-operon	Phenylacetic acid and derivatives	*Gordonia* sp. CWB2, *R. opacus* 1CP, *P. fluorescens* ST, *S. fribergensis* Kp5.2, recombinant	Oelschlägel et al. (2015a, b)
SMO, host cell	Ibuprofen	*Gordonia* sp. CWB2	Oelschlägel et al. (2014b, 2015b)

E1 (StyA/B) and E2 (StyA1/A2B) are the two types of SMOs (Montersino et al. 2011) and in fewer cases the catalysts were designated SMO since they originate from a metagenome. Besides StyB or StyA2B also (electro)chemical FAD reduction was performed. NDDH, naphthalene dihydrodiol dehydrogenase

be applied, several attempts can yield improved biocatalysts. Thus various host strains (*E. coli, Pseudomonas*) and respective carbon sources or vital state of cells were analyzed (Julsing et al. 2012; Kuhn et al. 2012a, b, 2013). Since pseudomonads seemed to be more resistant to toxic and cell integrity affecting compounds as styrene or styrene oxide these are favored in terms of stability. But, recombinant *E. coli* has been shown to provide higher biocatalytic activities and yields. That is probably caused due to the easier tunable growth behavior, simple handling of the host, and higher gene expression levels. Respectively, this host is favored in terms of productivity. In addition, whole cells producing SMO activity were combined with in vivo NADH regeneration systems as well as with further biocatalysts to produce more stable, longer, efficiently, and also other compounds then styrene oxide (Di Gennaro et al. 2013; Kuhn et al. 2013; McKenna et al. 2013; Oelschlägel et al. 2015a; Tischler et al. 2009; Toda et al. 2012a, b, 2014, 2015). Thus SMO in biotechnological application showed already a broad flexibility and are suitable in combinatorial approaches which is nowadays important, especially, in the field of synthetic biology. The latter is an emerging branch in the field and a bright future for biotechnology can be expected. Possible substrates and thus products of SMO-based oxidations had also been investigated and now an even more broad range of compounds can be produced (Table 5.1). Thus, very recently the efficient sulfoxidation of aromatic sulfides was demonstrated (Boyd et al. 2012; Nikodinovic-Runic et al. 2013; Paul et al. 2015). These reactions are also regio- and stereoselective and comparable in yields to conventional epoxidations performed by SMOs. However, the enantioselectivity of products obtained is often not as high as with epoxides produced. Thus, here directed evolution can be a route to evolve SMOs toward highly stereoselective sulfoxidation catalysts. First indications about amino acids involved in substrate binding (small versus large) and stereoselectivity were provided by Lin et al. (2012) and can be used to further understand the SMO catalysis and involved residues. Furthermore, the StyA1 protein (an E2 type SMO) had been shown to be highly enantioselective in epoxidation as well as sulfoxidation (Paul et al. 2015; Tischler et al. 2010). Thus it seems valuable to investigate more related proteins for their enantioselectivity in the monooxidation of substrates.

Styrene monooxygenases were also applied in cell-free systems to produce chiral products. Since the work of Hollmann et al. (2003), SMOs were combined with various NADH regeneration systems for production of reduced FAD or even the direct FAD reduction by (electro)chemical systems was studied (reviewed by Montersino et al. 2011). Further progress was made and shows that SMO epoxidase subunits can be applied without the respective reductase StyB at similar or even better catalytic efficiency (Paul et al. 2015). The reductase component is hard to express and to stabilize in large quantities and always a bottleneck in cell-free catalysis. Thus the electrochemical-driven FAD reduction and supply to StyA is favored for those systems. Recently, a corresponding system and reactor setup was demonstrated (Ruinatscha et al. 2014). However, the overall epoxidation rate is by far lower as that of the natural system StyB/StyA which is supposed due to limitations of the reactor design and transfer rates of reduced FAD. But, replacing NADH by the nicotinamide cofactor mimic 1-benzyl-1,4-dihydronicotinamide (BNAH) to reduce FAD for epoxidation or sulfoxidation with StyA1 had been shown to be more efficient and productive as the natural system StyA2B/StyA1 (Paul et al. 2015). Hence, this system seems most applicable at first glance, but, the demand of stoichiometric amounts of BNAH for production needs to be mentioned. And at the moment when there is no regeneration system for BNAH is available, further investigations are necessary to create an efficient cell-free system.

5.2 Biotechnological Potential of Styrene Oxide Isomerase

High stability, activity, simple enrichment, and storability of SOIs offer the opportunity for an application in the production of fine chemicals like phenylacetaldehyde and its derivatives (see Sect. 4.2). Those products were used as flavors, fragrances, or building blocks for pharmaceutical industry (Hölderich and Barsnick 2001; Miyamoto et al. 2007; Oelschlägel et al. 2012b). However, inhibition by the product phenylacetaldehyde formed limits the application of SOI as biocatalyst in a certain range (Oelschlägel et al. 2012a, 2014a, b). Final amounts of 760 μmol (=76 mM) pure phenylacetaldehyde were obtained after incubation of 130 U partially purified SOI of *Rhodococcus opacus* 1CP in presence of about 1 mmol styrene oxide in an aqueous system (=100 mM). Product formation rates of about 12 μmol phenylacetaldehyde per applied unit SOI could be reached. Remarkably, a recent study on the improvement of the SOI stability by covalent immobilization on silica-based carriers could not significantly stabilize the enzyme toward the phenylacetaldehyde, but improved its stability toward organic phases (Oelschlägel et al. 2014a). Based on those results, application of the covalently immobilized SOI in presence of a suitable organic phase could improve the product formation rates up to 20.0 μmol product per applied unit SOI (Oelschlägel et al. 2014a). Compared to an immobilized SOI, native enzyme preparation led only to 9.8 μmol product per unit applied enzyme under the same conditions. SOI is also usable as a part of an enzyme cascade additionally containing styrene monooxygenase and phenylacetaldehyde dehydrogenase for the production of phenylacetic acids from styrene and derivatives (see Table 5.1 and Sect. 5.6). And more recently it was demonstrated the SOI can be produced by means of an *E. coli* based expression system (Oelschlägel et al. 2015a). Thus now more of this biocatalyst can be produced without using toxic inducers as styrene or styrene oxide.

However, the substrate specificity of the styrene oxide isomerase was found to be rather limited (Miyamoto et al. 2007; Oelschlägel et al. 2012a, 2014b). Only styrene analogous compounds were found to be converted. The transmembrane character and absence of structural information would probably make it difficult to evolve SOIs in a direct fashion.

5.3 Phenylacetaldehyde Reductase in Biocatalysis

The activity of a phenylacetaldehyde reductase (PAR) has been reported from bacteria, fungi, and plant (Chen et al. 2011; Gopalakrishna et al. 1976; Hartmans et al. 1989; Itoh et al. 1997; Marconi et al. 1996; Tieman et al. 2007; Utkin et al. 1991). The systematic designation is 2-phenylethanol: NAD^+ oxidoreductase and a general overview has been given in Sect. 4.4. And it needs to be mentioned that in case of phenylacetaldehyde conversion 2-phenylethanol is produced, but the reverse reaction has not been observed. However, later it had been demonstrated that PAR is also active on alcohols as 2-propanol to produce the corresponding ketones (Itoh et al. 2002).

In terms of biotechnological application especially the enzyme from *Rhodococcus* sp. ST-10 (formerly designated as *Corynebacterium* sp. ST-10) has to be outlined (Itoh et al. 1997). The enzyme accepts solely NADH as source of reducing equivalents and has a rather broad substrate range (Itoh et al. 1997, 1999, 2002, 2007). Thus it allows the conversion of different prochiral 2-alkanones, aromatic ketones, and β-ketoesters to yield secondary alcohols with a high enantioselectivity (>99 %). Further, the enzyme is somewhat different in comparison to other alcohol dehydrogenases since it does not reduce acetone, acetaldehyde, or benzaldehyde (Itoh et al. 1997), but a number of other aliphatic and aromatic aldehydes. Respectively, it has a 12-fold higher activity on *n*-hexyl aldehyde as on the catabolic relevant phenylacetaldehyde. Ketones as, for example, 2-heptanone and some halogenated acetophenone derivatives are excellent substrates as well.

In order to achieve a competitive biotransformation and continuous product formation, a NADH-recycling system has to be coupled. Here, various approaches as formate dehydrogenase or PAR itself serve well (Itoh et al. 1999, 2002). Further, the enzyme can act in two-phase systems to overcome solubility issues of some substrates. Thus it is possible to employ PAR of strain ST-10 to produce building blocks for pharmaceutical industry at high yields: 28 g l^{-1} (*R*)-2-chloro-1-(3-chlorophenyl)ethanol from *m*-chlorophenacyl chloride, 28 g l^{-1} ethyl-(*R*)-4-chloro-3-hydroxy butanoate from ethyl-4-chloro-3-oxobutanoate, and 51 g l^{-1} (*S*)-*n*-tert-butoxycarbonyl(Boc)-3-pyrrolidinol from *n*-Boc-3-pyrrolidinone (Itoh et al. 2002). In most of these studies, the biocatalyst has been used in form of resting cells after expression.

By means of several mutagenesis studies, the PAR of strain ST-10 was optimized to overcome several bottlenecks on the route to an industrial relevant biocatalyst (Itoh et al. 2007; Makino et al. 2005, 2007). Since PAR allows product formation and in situ NADH regeneration, the rate-limiting step of that system had to be optimized first. Thus a mutant library was generated to allow functionality at high 2-propanol concentrations (about 20 % vol/vol) for a superior NADH production. This high 2-propanol concentration is also useful in order to allow the conversion of water-unsoluble substrates. And it had no effect on the substrate specificity or selectivity of PAR. Further optimization steps aimed to increase process stability and maintain or increase the substrate range (Makino et al. 2005, 2007). Thus it was possible to change PAR to accept also higher substrate concentrations and allow almost complete conversions (Makino and Itoh 2014) which was shown for the substrates *m*-chlorophenacyl chloride (2.1 mmol ml^{-1}) and ethyl-4-chloro-3-oxobutanoate (1.9 mmol ml^{-1}).

5.4 Epoxide Hydrolases Produce Chiral Diols or Remain Chiral Epoxides

Racemic epoxides can be converted in an enantioselective manner by epoxide hydrolases (EHs) which can yield chiral diols as a product or just act on a single stereoisomer of the racemic substrate while leaving a chiral epoxide behind which is known as kinetic resolution (Archer 1997; Breuer et al. 2004; Kotik et al. 2012; Orru and Faber 1999). For the mechanistic details of EHs and their metabolic role see Sect. 4.5. Racemic styrene oxide is the model substrate for EHs in order to evaluate productivity and selectivity. However, the autohydrolysis of epoxides has to be taken into account and to overcome the unfavored autohydrolysis a slight basic pH and a two-phase system can be applied (Liu et al. 2006). Fungal EHs had been extensively studied since they can be produced in large quantities and show a considerable enantioselectivity. A most prominent product obtained by enantiose-lective hydrolysis is (1S,2R)-indene oxide from racemic indene oxide which can be achieved by a fungal EH (Archer 1997; Breuer et al. 2004). This product is a building block for indinavir which is also known as crixivan and acts as a HIV protease inhibitor.

Epoxide hydrolases are studied already for decades and often a focus was put on application. Respectively, the source can be different and mammalian, plant, fungal, yeast, or bacterial enzymes have different properties which are beneficial for various biotransformations (reviewed by Kotik et al. 2012; Orru and Faber 1999). Often EHs are produced by recombinant gene expression in *E. coli* and crude extracts or whole cells (wild-type or recombinant) are applied for the bio-transformation. Rarely, the purified enzymes get employed, but if so it often was previously immobilized to stabilize the enzyme for a continuous process. In view cases enzymes had been optimized by (random) mutagenesis (Rui et al. 2004, 2005; van Loo et al. 2004; Zheng and Reetz 2010).

Many enzymes produce (S)-styrene oxide from racemic styrene oxide (Lee et al. 2004; Liu et al. 2006; Kim et al. 2006). Exemplary the EH originating from *Rhodotorula glutinis* was studied from the wild-type organism but also after recombinant production in *Pichia pastoris* (Lee et al. 2004; Weijers 1997). The recombinant version was about 10 times more active on the substrate (R)-styrene oxide (about 360 nmol min^{-1} per mg cell). Thus the product is a diol and (S)-styrene oxide remained. The yeast cells could be applied in a two-phase system while the racemic styrene oxide represented the organic phase. After 16 h, the kinetic resolution of 526 mM substrate resulted in 36 % yield (50 % theoreti-cal possible) at an enantiomeric excess of about 98 % (Lee et al. 2004). Similar enantioselectives had been obtained with other EHs as well origination exem-plarly form *Agrobacterium radiobacter* (Lutje Spelberg et al. 1998), *Aspergillus niger* strains (Choi et al. 1998; Pedragosa-Moreau et al. 1993, 1994), rabbit liver (Bellucci et al. 1993), *Rhodosporidium kratochvilovae* (Lee et al. 2003), *Sphingomonas echinoides* EH-983 (Kim et al. 2006), and *Sphingomonas* sp. HXN-200 (Liu et al. 2006). However, it is also possible to enrich the other epoxide enantiomer or enantiopure diols by employing the proper biocatalyst (Archelas and Furstoss 1997; Breuer et al. 2004; Manoj et al. 2001; Pedragosa-Moreau et al. 1993, 1994, 1996a, b). As indicated in Sect. 4.5, styrene oxides represent a special substrate group since both carbon atoms of the oxirane moiety can be attacked by the nucleophile generated from H_2O in EH active site (Breuer et al. 2004; Orru and Faber 1999; Pedragosa-Moreau et al. 1994, 1996b). EH of *A. niger* facilitates the H_2O-attack at C1-position of racemic *p*-chlorostyrene oxide to yield the (*R*)-diol (retention of stereochemistry) and remains the (*S*)-styrene oxide. The same diol is obtained by means of EH-activity from *Solanum tuberosum* or from *Beauveria sulfurescens* from kinetic resolution of racemic *p*-chlorostyrene oxide which means the attack of H_2O occurs at C2-position (inversion of stereochemistry). Hence, the (*R*)-styrene oxide gets enriched here. However, by combining both EH-activities, the racemic epoxide can be almost completely converted into a vicinal chiral diol. This enantiopure diol produced by both EHs can be applied to synthesize eliprodil a potential durg (Manoj et al. 2001).

Most of the above-mentioned EHs have a rather broad substrate spectrum and allow the kinetic resolution of styrene derivatives (Kotik et al. 2012) but also other epoxides. Here, especially epichlorohydrin has to be mentioned (Archelas and Furstoss 1997; Kim et al. 2006). The product of hydrolysis can be relevant for organic synthesis, but, is also carcinogen determined in food since the racemic epichlorohydrin is widely used in plastics, resins, and glues.

5.5 Other Enzymes Acting on Styrene and Its Derivatives

A number of enzymes not involved in styrene metabolism are able to transform styrene or styrene derivatives into valuable products (Fig. 5.1). Therefore, these enzymes might replace or act in concert with those of the *sty*-operon and provide new avenues to biotechnological applications.

The microbial oxidation of 2-bromostyrene by means of whole cell biocatalysis yielded (1*S*,2*R*)-4-bromo-3-ethenylcyclohexa-3,5-diene-1,2-diol and/or (1*R*)-1-(2-bromophenyl)ethan-1,2-diol (Fig. 5.1) (Königsberger and Hudlicky 1993). In the first case, the aromatic nucleus while in the second case the vinyl side-chain has been attacked. However, the products formed showed a high optical purity (>91 %) and the vinyl side-chain attack was favored. This might indicate an activity of a modified styrene degradative pathway as discussed in Sects. 2.2.4 and 2.4 which needs to be verified. In case of *Pseudomonas putida* 39/D, a mix of both products was obtained, whereas *Pseudomonas* sp. NCIB 9816-11 only produced (1*R*)-1-(2-bromophenyl)ethan-1,2-diol (Königsberger and Hudlicky 1993).

Variants of the cytochrome P450-monooxygenase BM3 and related enzymes were found to convert styrene and chemical analogous compounds in different manner (reviewed by Lechner et al. 2015). Thus. P450-like monooxygenases have been found to convert styrene selectively into styrene oxide (Cox et al. 1996; Fruetel et al. 1992, 1994; Li et al. 2001, 2013). Molecular oxygen gets activated at

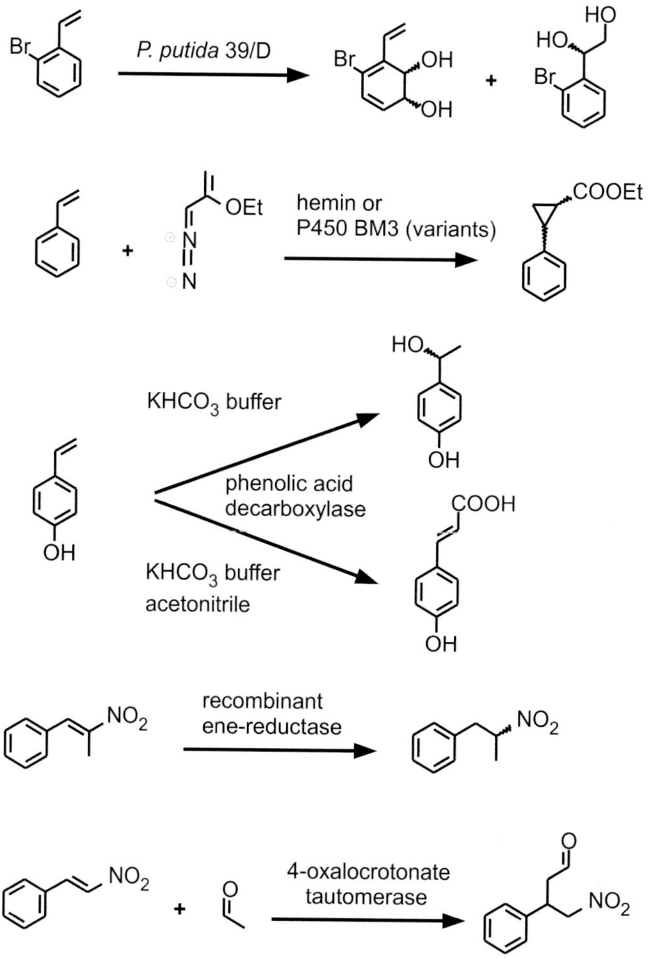

Fig. 5.1 Biotransformation of styrene (derivatives) is shown exemplary for the simplest substrates reported (Burda et al. 2013; Coelho et al. 2013a, b; Königsberger and Hudlicky 1993; Narancic et al. 2013; Wuensch et al. 2013, 2015) while often further derivatives can be converted. Potential chiral centers are indicated by flexible bonds

the heme center of the enzyme and allows the substrate monooxygenation which is in case of P450-monooxygenases often rather unselective and so hydroxylation, epoxidation, and sulfoxidations can be observed in dependence of the substrate. Styrene and its halogenated derivatives had been converted in a certain enantioselective manner to styrene oxide or derivatives. Thus, P450 Cam from *P. putida* and P450 pyr from *Sphingomonas* sp. HXN-20 yielded the (*S*)-enantiomer (66 % ee) while P450 BM3 from *Bacillus megaterium*, P450 pyr (triple mutant) from *Sphingomonas* sp. HXN-20, and P450 Terp from *Pseudomonas* sp. yielded the (R)-enantiomer (82 % ee; mutants up to 98 % ee). The P450 BM3 could be changed by single point mutation into an (R)-enantiomer producing biocatalyst and especially with 3-chlorostyrene as substrate 95 % of the (R)-enantiomer had been produced (Li et al. 2001).

Variants of the cytochrome P450 BM3 monooxygenase have been identified to use diazoester reagents as precursors in a formal carbene transfer onto styrene or p-substituted styrene derivatives (Coelho et al. 2013a). Therewith, the diastereoselective and enantioselective cyclopropanation of styrenes was demonstrated (Fig. 5.1). The mechanism of this cyclopropanation is comparable to the monooxygenation mentioned above. Here the diazoester gets coupled to the heme system forming the carbene precursor and releasing nitrogen. This heme-aduct allows the carbene transfer on styrene-like compounds. Even hemin as catalyst can be used but does only provide access to a racemic product of (R,R) and (S,S) cyclopropane derivatives. The enzymatic system achieves a better enantiomeric excess and by mutagenetic approaches enzymes can be obtained to produce further stereoisomers (Coelho et al. 2013a). This system can be further tuned to exclude monooxygenation activity by a single point mutation (Cys400Ser) (Coelho et al. 2013b). This variant has been designated 'P411' according to its distinct absorbance signature. Respectively, almost all variants reported had a higher activity as the free hemin and allowed to gain enriched isomers. But, in general, the total turnover numbers and conversion rates are rather low. Further, mutations had been introduced to enlarge the substrate range (Renata et al. 2014; Wang et al. 2014). These approaches have been carried out successfully and much bulkier substrates could be converted at better rates and with higher yields.

Styrene and derivatives as well as cyclic styrene derivatives as indene can be converted by peroxygenases (Kluge et al. 2012). Moderate turnover rates and enantioselectivities had been observed when applying a heme-thiolate peroxygenase originating from *Agrocybe aegerita*. Especially, the production of (1R,2S)-cis-β-methylstyrene oxide seems promising with an excellent enantiomeric excess with >99 % and a high total turnover number. The mechanism is, in general, similar to epoxidation reactions performed by heme-monooxygenases. But, here hydrogen peroxide serves as the source of oxygen which is activated at the heme center and transferred onto the substrate.

Styrene and its derivatives are also epoxidized by peroxidases (Colonna et al. 1993; Santhanam and Dordick 2002). Here, the chloroperoxidase from *Caldariomyces fumago* need to be mentioned. It produces mainly the (R)-enantiomer of styrene oxide with rather poor ee values of about 28–68 %. The oxygen is obtained from *tert*-butyl hydroperoxide and it is supposed that the oxygen is activated at the heme center and from the ferryl intermediate directly transferred onto the substrate. With respect to a biotransformation, this enzyme is unfavored since it provides only low epoxidation rates yielding little enantioenriched products, and even worse is its high catalase activity (Lechner et al. 2015).

The xylene monooxygenase from *P. putida* allows converting styrene and chemical analogous compounds into corresponding epoxides with moderate enantioselectivities (Panke et al. 1999; Wubbolts et al. 1994a, b). The enzyme can be produced by recombinant expression and the biomass obtained can be applied in form of resting cells for the biotransformation. Panke et al. (1999) constructed an efficient expression system inducible by octane, for example, and allowed so to produce enough enzyme for the biotransformation (91 U per g cell dry weight). The biocatalyst was then applied in a two-phase system composed of the medium and hexadecane and a high productivity of about 1 g styrene oxide per h and per liter aqueous phase has been accomplished (90 % conversion).

The *p*-hydroxystyrene and its derivatives are also known as *p*-vinylphenolic compounds. These compounds can be produced by chemical means or biotechnological approaches (see Sect. 5.6.4). The *p*-hydroxystyrene derivatives are an interesting class of precursors for hydratases (Wuensch et al. 2013). Thus it was surprisingly to identify a phenolic acid decarboxylase acting on *p*-hydroxystyrene-like substrates as a stereoselective hydratase yielding 1-(*p*-hydroxyphenyl)ethanol derivatives (Fig. 5.1). The degree of enantioselectivity obtained was found to be dependent on the bicarbonate buffer providing a chance to enlight the enzyme mechanism but also tuning the reaction. Seven enzymes originating from various microorganisms had been analyzed and showed different conversion and stereoselectivites on various substrates (Wuensch et al. 2013). Respectively, a novel route to enantio-enriched 1-phenylethanol-like compounds had been described.

Interestingly, the same phenolic acid decarboxylase has been found to convert *p*-hydroxystyrene derivatives into (*E*)-cinnamic acid derivatives (Fig. 5.1) (Wuensch et al. 2015). The *p*-hydroxy group has to be found to be mandatory for the conversion. Bicarbonate serves as the carbon dioxide source in the β-caboxylation, while a high regio- and stereoselectivity was observed. The degree of substitutes at the aromatic nucleus negatively effects the conversion, but substitutions at the vinyl side-chain yielded almost no product at all. However, this can be an interesting strategy to yield cinnamic acid-like compounds.

Ene-reductases also can act on styrene derivatives if they are activated, for example, by an adjacent nitro-group (Fig. 5.1) (Burda et al. 2013). The products obtained are important building blocks for chirale amines in order to produce pharmaceuticals as exemplary tamsulosin or selegiline. Gröger and coworkers used a recombinant ene-reductase originating from *Gluconobacter oxydans* to reduce α-methylated *trans*-nitroalkenes (Burda et al. 2013). The substrate and the reaction had to be emulsified or mixed by ultrasound. NADPH served as source of reducing equivalents and had to be recycled in order to achieve conversions. The reaction was improved by process optimization (e.g., temperature, mixing, and timing) to achieve a higher enantioselectivity (up 95 % ee) in case of (*R*)-nitroalkane production. Substitutions at the aromatic systems had, in most cases, no drastic effect on conversion and stereoselectivity.

Similar as the ene-reductases also a 4-oxalocrotonate tautomerase can act on a nitro-activated styrene-like compound (Narancic et al. 2013). Here, β-nitrostyrene and acetaldehyde are the substrates of a Michael-type addition (Fig. 5.1), while the simplest product 4-nitro-3-phenyl-butanol had been obtained at high yields (60 %) and with an excellent enantioselectivity (>99 % ee). The enzyme was recombinantly produced and the resting cells harboring the biocatalyst had been applied to perform the biotransformation. A number of substrate derivatives had been tested and could be converted at lower yields and with lower selectivity.

5.6 Production of Valuable Compounds Employing Enzymatic Cascades

In order to produce or to enrich intermediates or derivatives of pathways involved in styrene catabolism or to combine these with other enzymes various approaches have been reported (see also Table 5.1).

5.6.1 Styrene-Degrading Bacteria Produce Phenylacetic Acids

As already mentioned, styrene-degrading bacteria are able to degrade styrene via styrene oxide and phenylacetaldehyde to phenylacetic acid. For that reaction, the enzymes SMO, SOI, and PAD act together in an enzyme cascade which also recycles all the cofactors which are necessary for these steps (Fig. 5.2). It has recently

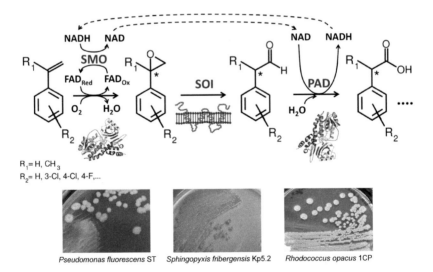

Fig. 5.2 The enzymatic cascade of the upper styrene degradation pathway employed to produced phenylacetic acid and its derivatives. Together, the enzymes SMO, SOI, and PAD are cofactor independent as indicated by the internal recycling. Three soil bacteria possessing this route are shown below the cascade. Strain CWB2 is supposed to utilize a different pathway and thus it is not highlighted here

been shown that also substituted and halogenated styrenes are converted into the corresponding phenylacetic acids (Oelschlägel et al. 2015b). In contrast to the non-substituted phenylacetic acids, especially the halogenated products are not further metabolized during the phenylacetic acid degradation (Oelschlägel et al. 2015b). These phenylacetic acid derivatives formed are secreted into the culture medium and can be enriched from it to practicable concentrations. In the study of Oelschlägel et al. (2015b), the styrene-degrading strains *Pseudomonas fluorescens* ST, *Sphingopyxis fribergensis* Kp5.2, *R. opacus* 1CP, and *Gordonia* sp. CWB2 were investigated for the production of different substituted phenylacetic acids. It was shown that the soil bacteria differ significantly in their substrate tolerance and the reachable yields of products. In summary, especially *P. fluorescens* ST has been identified as promising biocatalyst for the high-selective production of such products. The whole-cell biotransformation of 4-chlorostyrene by strain ST yielded about 4.7 g l^{-1} product after nearly 350 days (Oelschlägel et al. 2015b). This result shows the remarkable runtime of this system and gives leeway for significant optimization.

While more bulky substrates were not transformed into the corresponding acids by the strains ST, Kp5.2, and 1CP, the isolate *Gordonia* sp. CWB2 was able to catalyze the reaction of 4-isobutyl-α-methylstyrene into the pharmaceutical 4-isobutyl-α-methylphenylacetic acid (Oelschlägel et al. 2015b). This product is better known as ibuprofen. The extended substrate spectrum in the case of strain CWB2 indicates differences in the enzymes involved in the cascade. A previous study has reported on SOI activity from styrene-grown biomass of strains Kp5.2, 1CP, and ST (Beltrametti et al. 1997; Bestetti et al. 2004; Oelschlägel et al. 2014b). Furthermore, the corresponding SOI-genes were found on the genome of these organisms together with genes encoding for the SMO and the PAD. Remarkably, attempts to detect the *styC*-encoded styrene oxide isomerase in case of strain CWB2 failed because no corresponding activity was found (Oelschlägel et al. 2014b). These results strongly indicate that strain CWB2 has a modified way of side-chain oxygenation yielding phenylacetic acids from styrenes. This modification enables this strain to transform more bulky substrates which makes this strain to a promising catalyst for the production of such pharmaceuticals as ibuprofen.

5.6.2 Production of Phenylacetic Acids by Recombinant Biocatalysts

During a more recent study, a recombinant biocatalyst was constructed based on *E. coli* BL21 (DE3) as host in order to produce phenylacetic acids from styrene oxides (Oelschlägel et al. 2015a). For that, the gene sequence information of the SOI of *R. opacus* 1CP was adjusted to the codon usage of the host strain, synthesized by gene-synthesis, and the gene subsequently transformed in strain BL21 (DE3). Because a further gene designated as *feaB* is already present in the genome of this host strain and encodes for a phenylacetaldehyde dehydrogenase (PAD, or FeaB), the simple modification of strain BL21 (DE3) by the SOI gene addition enables the recombinant microorganism to produce up to 0.73 g l^{-1} non-substituted phenylacetic acid (a yield of about 85 %) from styrene oxide over a period of 20 days (Oelschlägel et al. 2015a). In contrast to styrene-degrading wild-type strains, this recombinant strain enables also the production of non-substituted phenylacetic acid beside substituted products from the corresponding styrene oxide derivatives because *E. coli* BL21 (DE3) harbors no gene cluster encoding enzymes of the phenylacetic acid degradation (Oelschlägel et al. 2015a). The additional transformation of an SMO gene to this *E. coli* strain mentioned above may provide a whole-cell biocatalyst which is usable for the complete conversion of styrene and its derivatives into phenylacetic acid and related compounds similar as known for wild-type strains (Sect. 5.6.1).

It can be summarized that not only the single enzymes of the styrene degradation pathway of the side-chain oxygenation are of interest for biotechnological applications but also the complete pathway can be used in order to synthesize pharmaceuticals and other fine chemicals with industrial relevance.

5.6.3 Biosynthesis of Catechol Derivatives with Rhodococcus

The strain *Rhodococcus rhodochrous* NCIMB 13259 degrades styrene via a direct attack at the aromatic system and a later *meta*-cleavage route as described earlier (Warhurst et al. 1994). During this biotransformation, 3-vinylcatechol was determined and by altering the biological system its production could be achieved. The activity of the here 3-vinylcatechol converting extra-diol dioxygenase was inhibited by means of 3-fluorocatechol which is a general inhibitor of *meta*-cleaving dioxygenases (Mars et al. 1997). Thus only the first two enzymes of the pathway act as a cascade: styrene dioxygenase and dihydrodiol reductase. Furthermore, due to their relaxed substrate specificity also toluene and ethylbenzene had been converted to 3-methylcatechol as well as to 3-ethylcatechol. Such compounds are interesting building blocks for polymers and pharmaceuticals and often hardly available by conventional syntheses. It might be a valuable target to check if other styrene degraders (Oelschlägel et al. 2014b) are also able to produce catechol-like compounds and compare yields and rates, respectively.

5.6.4 Production of 4-Hydroxystyrene

For the production of polymers, *p*-HO styrene (4-hydroxystyrene) is an interesting monomer. Furthermore, it might serve as a substrate for a styrene-transforming cascade to yield valuable (aroma) compounds as hydroxylated derivatives of phenylacetic acid or 2-phenylethanol (see Sects. 5.6.1 and 5.6.2). Thus a biotechnological production of *p*-HO styrene had been investigated (Qi et al. 2007; Verhoef et al. 2009). By means of molecular genetic approaches, two enzymes had first been introduced to *E. coli* WWQ51.1 and later into *P. putida* strain S12 in order to allow the product formation starting from glucose as a carbon source growing again. The central intermediate tyrosine can be produced by the host and due to the activity of a tyrosine ammonia lyase originating from *R. glutinis* and a *p*-coumaric acid decarboxylase originating from *Lactobacillus plantarum* *p*-HO styrene is produced. The later used pseudomonad is just more tolerant to organic solvents and thus two-phase systems get applicable and the product yield could be increased (Verhoef et al. 2009). The process awaits transition to a pilot scale.

Similar systems had been evolved to overproduce other monomers besides the 4-hydroxystyrene as exemplary styrene itself and the 3,4-dihydroxystyrene as well as the 4-hydroxy-3-methoxystyrene (Kang et al. 2015; McKenna et al. 2014). The following titers of products had been determined: 29–260 mg l^{-1} styrene, 355 mg l^{-1} 4-hydroxystyrene, 63 mg l^{-1} 3,4-dihydroxystyrene, and 64 mg l^{-1} 4-hydroxy-3-methoxystyrene. And it has to be mentioned that the renewable styrene production was achieved by either an evolved *E. coli* or an evolved *Saccharomyces cerevisiae* strain while all the other studies only used *E. coli* as model-organism (McKenna and Nielsen 2011; McKenna et al. 2014).

Another route toward 4-hydroxystyrene and its derivatives is the selective vinylation of phenols with pyruvate (Busto et al. 2015). Here, a three-step enzymatic cascade gets employed which is composed of a tyrosine phenol lyase, a tyrosine ammonia lyase, and a ferulic acid decarboxylase (Fig. 5.3). In this process, ammonia gets recycled. Further, only water and carbon dioxide are the by-products. The process accepts a number of phenol derivatives and is highly regioselective since only *p*-vinylation occurs.

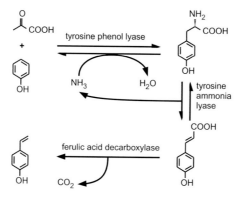

Fig. 5.3   Enzymatic cascade to produce 4-hydroxystyrene (Busto et al. 2015)

5.7 Applying Biomass for Waste Treatment

In order to remove volatile organic compounds (VOCs) from waste gas streams conventional (active carbon filtering, combustion processes) or biological (bio-filters, bio-trickling filters, bioscrubbers) methods can be applied (Delhomenie and Heitz 2005; Deshusses 1997; Malhautier et al. 2005). The biological vari-ants are often cost-effective and a natural microbial flora can be employed to rapidly degrade VOCs (Arnold et al. 1997; Corsi and Seed 1995; Juneson et al. 2001). But, it needs to be mentioned that the VOC concentration generally need to be low enough to achieve an almost complete removal prior the gas stream enters the atmosphere. Readily bioavailable but also harmful compounds as sty-rene are model substances to set up and investigate VOC removal by means of biological systems. Often bio-trickling filters (Fig. 5.4) are employed to degrade VOCs rapidly. Here a certain packing material which can be an inert matrix or also of biological origin as wood particles can serve as the support for the microorganisms. These can be stimulated by an online nutrient supply or just humidified to achieve a degradative potential. An almost natural variant of such a biological system to treat waste streams from a German company is outlined in Sect. 6.2 and a detailed overview on consortia or (developed) strains in order to remove VOCs containing also styrene was presented earlier (Tischler and Kaschabek 2012).

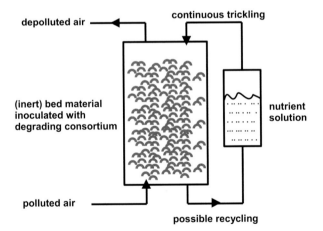

Fig. 5.4 The general structure of a bio-trickling filter is shown. The bed material can be inert or biological material to support the microbial growth usually observed as biofilm. The biomass can be stimulated by adding nutrients or just humidified in order to achieve a certain biological activity

5.8 Biopolymer Production

The styrene degrader *P. putida* CA-3 produces polyhydroxyalkanoates (PHAs) under nitrogen limiting conditions upon styrene consumption (Nikodinovic-Runic et al. 2009; O'Connor et al. 1996; Ward et al. 2005). The product obtained had been analyzed and comprises the following building blocks: 3-hydroxyhexanoate, 3-hydroxyoctanoate, and 3-hydroxydocanoate in a ratio of 3:22:75. Strain CA-3 was shown to yield most PHA per utilized substrate in comparison to other *Pseudomonas* strains and a yield of about 23 % of cell-dry weight had been determined. PHA production increased with the cultivation time at low nitrogen levels (about 2–3 mg l^{-1}) (Nikodinovic-Runic et al. 2009). And from those experiments, PHA biosynthetic relevant enzymes had been determined by means of a proteomic approach. In comparison to the styrene-degrading enzymes, those proteins were present at low levels and any PHA depolymerase was even not detected. Interestingly, polystyrene was chosen as starting material to produce PHAs by means of strain CA-3 (Ward et al. 2006). However, polystyrene is poorly bioavailable (Mor and Sivan 2008) and likely the enzymes involved in the styrene catabolism are not involved. To overcome this limitation, polystyrene can be thermally pre-treated to yield the monomer which is used as feedstock (Ward et al. 2006). However, this might be a valuable idea to make use of the recalcitrant polymer, but, one has to consider the high energy demand and the low PHA-yield of strain CA-3.

5.9 Biosensors

Occurrence of numerous toxic and recalcitrant organics such as BTEX in nature by anthropogenic activities has led to the development of diverse methods for in situ determination of respective compounds. Thus also biological sensors (biosensors) were developed especially to measure bioavailability of hydrocarbons which may contaminate soil, water, or air. Usually, a combination of a genetic regulatory element (e.g., *xylR*, *xylS*, *tod*) and a reporter system (e.g., *gfp*, *luxAB*, *lacZ*) can be used for the detection of contaminants (Keane et al. 2002; Rodriguez-Mozaz et al. 2006; Stiner and Halverson 2002).

The *sty*-operon is often regulated by a StyS-StyR system which sense for styrene and by a positive response induces the expression of further *sty*-genes (see Sect. 3.2). Thus *styR* was used as a regulatory element to determine the presence of styrene and respectively induces the expression of a *beta*-galactosidase which acts as reporter (Alonso et al. 2003). As usual, the system constructed was found to be sensitive only for styrene since it accepts also other compounds as phenylacetaldehyde, 2-phenylethanol, styrene oxide, and toluene. A similar approach by

Stiner and Halverson (2002) based on the toluene-benzene utilization gene cluster and its regulatory elements which were combined with a green fluorescent protein as reporter. The systems worked quantitatively for number of hydrocarbons including styrene.

References

Alonso S, Navarro-Llorens JM, Tormo A, Perera J (2003) Construction of a bacterial biosensor for styrene. J Biotechnol 102:301–306
Archelas A, Furstoss R (1997) Synthesis of enantiopure epoxides through biocatalytic approaches. Annu Rev Microbiol 51:491–525
Archer IVJ (1997) Epoxide hydrolases as asymmetric catalysts. Tetrahedron 53:15617–15662
Arnold M, Reittu A, Von Wright A, Martikainen PJ, Suihko ML (1997) Bacterial degradation of styrene in waste gases using a peat filter. Appl Microbiol Biotechnol 48:738–744
Bellucci G, Chiappe C, Cordoni A, Marioni F (1993) Substrate enantioselectivity in the rabbit liver microsomal epoxide hydrolase catalyzed hydrolysis of *trans* and *cis* 1-phenylpropene oxides. A comparison with styrene oxide. Tetrahedron Asymmetry 4:1153–1160
Beltrametti F, Marconi AM, Bestetti G, Galli E, Ruzzi M, Zennaro E (1997) Sequencing and functional analysis of styrene catabolism genes from *Pseudomonas fluorescens* ST. Appl Environ Microbiol 63:2232–2239
Bernasconi S, Orsini F, Sello G, Colmegna A, Galli E, Bestetti G (2000) Bioconversion of substituted styrenes to the corresponding enantiomerically pure epoxides by a recombinant *Escherichia coli* strain. Tetrahedron Lett 41:9157–9161
Bernasconi S, Orsini F, Sello G, Di Gennaro P (2004) Bacterial monooxygenase mediated preparation of chiral oxiranes: study of the effects of substituent nature and position. Tetrahedron Asymmetry 15:1603–1606
Bestetti G, Di Gennaro P, Colmegna A, Ronco I, Galli E, Sello G (2004) Characterization of styrene catabolic pathway in *Pseudomonas fluorescens* ST. Int Biodet Biodeg 54:183–187
Boyd DR, Sharma ND, McMurray B, Haughey SA, Allen CCR, Hamilton JTG, McRoberts WC, O'Ferrall RAM, Nikodinovic-Runic J, Coulombel LA, O'Connor KE (2012) Bacterial dioxygenase- and monooxygenase-catalysed sulfoxidation of benzo[*b*]thiophenes. Org Biomol Chem 10:782–790
Breuer M, Ditrich K, Habicher T, Hauer B, Kesseler M, Stuermer R, Zelinski T (2004) Industrial methods for the production of optically active intermediates. Angew Chem Int Ed 43:788–824
Burda E, Reß T, Winkler T, Giese C, Kostrov X, Huber T, Hummel W, Gröger H (2013) Highly enantioselective reduction of α-methylated nitroalkenes. Angew Chem Int Ed 52:9323–9326
Busto E, Simon RC, Kroutil W (2015) Vinylation of unprotected phenols using a biocatalytic system. Angew Chem Int Ed. doi:10.1002/anie.201505696
Chen X-M, Kobayashi H, Sakai M, Hirata H, Asai T, Ohnishi T, Baldermann S, Watanabe N (2011) Functional characterization of rose phenylacetaldehyde reductase (PAR), an enzyme involved in the biosynthesis of the scent compound 2-phenylethanol. J Plant Physiol 168:88–95
Choi WJ, Huh EC, Park HJ, Lee EY, Choi CY (1998) Kinetic resolution for optically active epoxides by microbial enantioselective hydrolysis. Biotechnol Tech 12:225–228
Coelho PS, Brustad EM, Kannan A, Arnold FH (2013a) Olefin cyclopropanation via carbene transfer catalyzed by engineered cytochrome P450 enzymes. Science 339:307–310
Coelho PS, Wang ZJ, Ener ME, Baril SA, Kannan A, Arnold FH, Brustad EM (2013b) A serine-substituted P450 catalyzes highly efficient carbene transfer to olefins in vivo. Nat Chem Biol 9:485–490

Colonna S, Gaggero N, Casella L, Carrea G, Pasta P (1993) Enantioselective epoxidation of styrene derivatives by chloroperoxidase catalysis. Tetrahedron Asymmetry 4:1325–1330

Corsi RL, Seed L (1995) Biofiltration of BTEX: media, substrate, and loadings effects. Environ Prog 14:151–158

Cox HHJ, Faber BW, van Heiningen WNM, Radhoe H, Doddema HJ, Harder W (1996) Styrene metabolism in *Exophiala jeanselmei* and involvement of a cytochrome P-450-dependent styrene monooxygenase. Appl Environ Microbiol 62:1471–1474

Delhomenie M-C, Heitz M (2005) Biofiltration of air: a review. Crit Rev Biotechnol 25:53–72

Deshusses MA (1997) Biological waste air treatment in biofilters. Curr Opin Biotechnol 8:335–339

Di Gennaro P, Colmegna A, Galli E, Sello G, Pelizzoni F, Bestetti G (1999) A new biocatalyst for production of optically pure aryl epoxides by styrene monooxygenase from *Pseudomonas fluorescens* ST. Appl Environ Microbiol 65:2794–2797

Di Gennaro P, Kazandjian LV, Mezzetti F, Sello G (2013) Regulated expression systems for the development of whole-cell biocatalysts expressing oxidative enzymes in a sequential manner. Arch Microbiol 195:269–278

Fruetel JA, Collins JR, Camper DL, Loew GH, Demontellano PRO (1992) Calculated and experimental absolute stereochemistry of the styrene and beta-methylstyrene epoxides formed by cytochrome-P450(Cam). J Am Chem Soc 114:6987–6993

Fruetel JA, Mackman RL, Peterson JA, Demontellano PRO (1994) Relationship of active-site topology to substrate-specificity for cytochrome P450(Terp) (Cyp108). J Biol Chem 269:28815–28821

Gopalakrishna Y, Narayanan TK, Ramanada Rao G (1976) Biosynthesis of β-phenethyl alcohol in *Candida guilliermondii*. Biochem Biophys Res Comm 69:417–422

Guan C, Ju J, Borlee BR, Williamson LL, Shen B, Raffa KF, Handelsman J (2007) Signal mimics derived from a metagenomic analysis of the gypsy moth gut microbiota. Appl Environ Microbiol 73:3669–3676

Gursky L, Nikodinovic-Runic J, Feenstra K, O'Connor K (2010) In vitro evolution of styrene monooxygenase from *Pseudomonas putida* CA-3 for improved epoxide synthesis. Appl Microbiol Biotechnol 85:995–1004

Hartmans S, Smits JP, van der Werf MJ, Volkering F, de Bont JAM (1989) Metabolism of styrene oxide and 2-phenylethanol in the styrene-degrading *Xanthobacter* strain 124X. Appl Environ Microbiol 55:2850–2855

Hölderich WH, Barsnick U (2001) Rearrangement of epoxides. In: Sheldon SA, van Bekkum H (eds) Fine chemicals through heterogeneous catalysis. Wiley-VCH, Weinheim, pp 217–231

Hollmann F, Lin P-C, Witholt B, Schmid A (2003) Stereospecific biocatalytic epoxidation: the first example of direct regeneration of a FAD-dependent monooxygenase for catalysis. J Am Chem Soc 125:8209–8217

Huijbers MME, Montersino S, Westphal AH, Tischler D, van Berkel WJH (2014) Flavin dependent monooxygenases. Arch Biochem Biophys 544:2–17

Itoh N, Morihama R, Wang J, Okada K, Mizuguchi N (1997) Purification and characterization of phenylacetaldehyde reductase from a styrene-assimilating *Corynebacterium* strain, ST-10. Appl Environ Microbiol 63:3783–3788

Itoh N, Mizuguchi N, Mabuchi M (1999) Production of chiral alcohols by enantioselective reduction with NADH-dependent phenylacetaldehyde reductase from *Corynebacterium* strain, ST-10. J Mol Catal B Enzym 6:41–50

Itoh N, Matsuda M, Mabuchi M, Dairi T, Wang J (2002) Chiral alcohol production by NADH-dependent phenylacetaldehyde reductase coupled with in situ regeneration of NADH. Eur J Biochem 269:2394–2402

Itoh N, Nakamura M, Inoue K, Makino Y (2007) Continuous production of chiral 1,3-butanediol using immobilized biocatalysts in a packed bed reactor: promising biocatalysis method with an asymmetric hydrogen-transfer bioreduction. Appl Microbiol Biotechnol 75:1249–1256

Julsing MK, Kuhn D, Schmid A, Bühler B (2012) Resting cells of recombinant *E. coli* show high epoxidation yields on energy source and high sensitivity to product inhibition. Biotechnol Bioeng 109:1109–1119

Juneson C, Ward OP, Singh A (2001) Microbial treatment of a styrene-contaminated air stream in a biofilter with high elimination capacities. J Ind Microbiol Biotechnol 26:196–202

Kang S-Y, Choi O, Lee JK, Ahn J-O, Ahn JS, Hwang BY, Hong Y-S (2015) Artificial de novo biosynthesis of hydroxystyrene derivatives in a tyrosine overproducing *Escherichia coli* strain. Microb Cell Fact 14:78

Keane A, Phoenix P, Ghoshal S, Lau PCK (2002) Exposing culprit organic pollutants: a review. J Microbiol Methods 49:103–119

Kim HS, Lee OK, Lee SJ, Hwang S, Kim SJ, Yang S-H, Park S, Lee EY (2006) Enantioselective epoxide hydrolase activity of a newly isolated microorganism, *Sphingomonas echinoides* EH-983, from seawater. J Mol Catal B Enzym 41:130–135

Kluge M, Ullrich R, Scheibner K, Hofrichter M (2012) Stereoselective benzylic hydroxylation of alkylbenzenes and epoxidation of styrene derivatives catalyzed by the peroxygenase of *Agrocybe aegerita*. Green Chem 14:440–446

Königsberger K, Hudlicky T (1993) Microbial oxidation of 2-bromostyrene by *Pseudomonas putida* 39/D. Isolation and identification of metabolites. Tetrahedron Asymmetry 4:2469–2474

Kotik M, Archelas A, Wohlgemuth R (2012) Epoxide hydrolases and their application in organic synthesis. Curr Org Chem 16:451–482

Kuhn D, Bühler B, Schmid A (2012a) Production host selection for asymmetric styrene epoxidation: *Escherichia coli* vs. solvent-tolerant *Pseudomonas*. J Ind Microbiol Biotechnol 39:1125–1133

Kuhn D, Julsing MK, Heinzle E, Bühler B (2012b) Systematic optimization of a biocatalytic two-liquid phase oxyfunctionalization process guided by ecological and economic assessment. Green Chem 14:645–653

Kuhn D, Fritzsch FSO, Zhang X, Wendisch VF, Blank LM, Bühler B, Schmid A (2013) Subtoxic product levels limit the epoxidation capacity of recombinant *E. coli* by increasing microbial energy demands. J Biotechnol 163:194–203

Lechner H, Pressnitz D, Kroutil W (2015) Biocatalysts for the formation of three- to six-membered carbo- and heterocycles. Biotechnol Adv 33:457–480

Lee JW, Lee EJ, Yoo SS, Park SH, Kim HS, Lee EY, Lee EY (2003) Enantioselective hydrolysis of racemic styrene oxide by epoxide hydrolase of *Rhodosporidium kratochvilovae* SYU-08. Biotechnol Bioprocess Eng 8:306–308

Lee EY, Yoo S-S, Kim HS, Lee SJ, Oh Y-K, Park S (2004) Production of (*S*)-styrene oxide by recombinant *Pichia pastoris* containing epoxide hydrolase from *Rhodotorula glutinis*. Enz Microbiol Technol 35:624–631

Li QS, Ogawa J, Schmid RD, Shimizu S (2001) Residue size at position 87 of cytochrome P450BM-3 determines its stereoselectivity in propylbenzene and 3-chlorostyrene oxidation. FEBS Lett 508:249–252

Li AT, Liu J, Pham SQ, Li Z (2013) Engineered P450pyr monooxygenase for asymmetric epoxidation of alkenes with unique and high enantioselectivity. Chem Commun 49:11572–11574

Lin H, Qiao J, Liu Y, Wu Z-L (2010) Styrene monooxygenase from *Pseudomonas* sp. LQ26 catalyzes the asymmetric epoxidation of both conjugated and unconjugated alkenes. J Mol Catal B Enzym 67:236–241

Lin H, Liu Y, Wu ZL (2011) Highly diastereo- and enantio-selective epoxidation of secondary allylic alcohols catalyzed by styrene monooxygenase. Chem Commun (Camb) 47:2610–2612

Lin H, Tang D-F, Qaed Ahmed AA, Liu Y, Wu Z-L (2012) Mutations at the putative active cavity of styrene monooxygenase: enhanced activity and reversed enantioselectivity. J Biotechnol 161:235–241

Liu Z, Michel J, Wang Z, Witholt B, Li Z (2006) Enantioselective hydrolysis of styrene oxide with the epoxide hydrolase of *Sphingomonas* sp. HXN-200. Tetrahedron Asymmetry 17:47–52

Makino Y, Itho N (2014) Development of an improved phenylacetaldehyde reductase mutant by an efficient selection procedure. Appl Microbiol Biotechnol 98:4437–4443

Makino Y, Inoue K, Dairi T, Itoh N (2005) Engineering of phenylacetaldehyde reductase for efficient substrate conversion in concentrated 2-propanol. Appl Environ Microbiol 71:4713–4720

Makino Y, Dairi T, Itoh N (2007) Engineering the phenylacetaldehyde reductase mutant for improved substrate conversion in the presence of concentrated 2-propanol. Appl Microbiol Biotechnol 77:833–843

Malhautier L, Khammar N, Bayle S, Fanlo J-L (2005) Biofiltration of volatile organic compounds. Appl Microbiol Biotechnol 68:16–22

Manoj KM, Archelas A, Baratti J, Furstoss R (2001) Microbiological transformations. Part 45. A green chemistry preparative scale synthesis of enantiopure building blocks of eliprodil: elaboration of a high substrate concentration epoxide hydrolase-catalyzed hydrolytic kinetic resolution process. Tetrahedron 57:695–701

Marconi AM, Beltrametti F, Bestetti G, Solinas F, Ruzzi M, Galli E, Zennaro E (1996) Cloning and characterization of styrene catabolism genes from *Pseudomonas fluorescens* ST. Appl Environ Microbiol 62:121–127

Mars AE, Kasberg T, Kaschabek SR, van Agteren MH, Janssen DB, Reineke W (1997) Microbial degradation of chloroaromatics: use of the *meta*-cleavage pathway for mineralization of chlorobenzene. J Bacteriol 179:4530–4537

McKenna R, Nielsen DR (2011) Styrene biosynthesis from glucose by engineered *E. coli*. Metab Eng 13:544–554

McKenna R, Pugh S, Thompson B, Nielsen DR (2013) Microbial production of the aromatic building-blocks (*S*)-styrene oxide and (*R*)-1,2-phenylethanediol from renewable resources. Biotechnol J 8:1465–1475

McKenna R, Thompson B, Pugh S, Nielsen DR (2014) Rational and combinatorial approaches to engineering styrene production by *Saccharomyces cerevisiae*. Microb Cell Fact 13:123

Miyamoto K, Okuro K, Ohta H (2007) Substrate specificity and reaction mechanism of recombinant styrene oxide isomerase from *Pseudomonas putida* S12. Tetrahedron Lett 48:3255–3257

Montersino S, Tischler D, Gassner GT, van Berkel WJH (2011) Catalytic and structural features of flavoprotein hydroxylases and epoxidases. Adv Synth Catal 353:2301–2319

Mor R, Sivan A (2008) Biofilm formation and partial biodegradation of polystyrene by the actinomycete *Rhodococcus ruber*: biodegradation of polystyrene. Biodegradation 19:851–858

Narancic T, Radivojevic J, Jovanovic P, Francuski D, Bigovic M, Maslak V, Savic V, Vasiljevic B, O'Connor KE, Nikodinovic-Runic J (2013) Highly efficient Michael-type addition of acetaldehyde by β-nitrostyrenes by whole resting cells of *Escherichia coli* expressing 4-oxalocrotonate tautomerase. Biores Technol 142:462–468

Nikodinovic-Runic J, Flanagan M, Hume AR, Cagney G, O'Connor KE (2009) Analysis of the *Pseudomonas putida* CA-3 proteome during growth on styrene under nitrogen-limiting and non-limiting conditions. Microbiology 155:3348–3361

Nikodinovic-Runic J, Coulombel LA, Francuski D, Sharma ND, Boyd DR, O'Ferrall RAM, O'Connor KE (2013) The oxidation of alkylaryl sulfides and benzo[*b*]thiophenes by *Escherichia coli* cells expressing wild-type and engineered styrene monooxygenase from *Pseudomonas putida* CA-3. Appl Microbiol Biotechnol 97:4849–4858

O'Connor K, Duetz W, Wind B, Dobson ADW (1996) The effect of nutrient limitation on styrene metabolism in *Pseudomonas putida* CA-3. Appl Environ Microbiol 62:3594–3599

O'Connor KE, Dobson AD, Hartmans S (1997) Indigo formation by microorganisms expressing styrene monooxygenase activity. Appl Environ Microbiol 63:4287–4291

Oelschlägel M, Gröning JAD, Tischler D, Kaschabek SR, Schlömann M (2012a) Styrene oxide isomerase of *Rhodococcus opacus* 1CP, a highly stable and considerably active enzyme. Appl Environ Microbiol 78:4330–4337

Oelschlägel M, Tischler D, Gröning JAD, Kaschabek SR, Schlömann M (2012b) Process for the enzymatic synthesis of aromatic aldehydes or ketones. Patent: DE 102011006459 A1 20121004

Oelschlägel M, Riedel A, Zniszczoł A, Szymańska K, Jarzębski AB, Schlömann M, Tischler D (2014a) Immobilization of an integral membrane protein for biotechnological phenylacetaldehyde production. J Biotechnol 174:7–13

Oelschlägel M, Zimmerling J, Schlömann M, Tischler D (2014b) Styrene oxide isomerase of *Sphingopyxis* sp. Kp5.2. Microbiol (UK) 160:2481–2491

Oelschlägel M, Heiland C, Schlömann M, Tischler D (2015a) Production of a recombinant membrane protein in an *Escherichia coli* strain for the whole cell biosynthesis of phenylacetic acids. Biotechnol Rep 7:38–43

Oelschlägel M, Kaschabek SR, Zimmerling J, Schlömann M, Tischler D (2015b) Co-metabolic formation of substituted phenylacetic acids by styrene-degrading bacteria. Biotechnol Rep 6:20–26

Orru RV, Faber K (1999) Stereoselectivities of microbial epoxide hydrolases. Curr Opin Chem Biol 3:16–21

Otto K, Hofstetter K, Roethlisberger M, Witholt B, Schmid A (2004) Biochemical characterization of StyAB from *Pseudomonas* sp. strain VLB120 as a two-component flavin-diffusible monooxygenase. J Bacteriol 186:5292–5302

Panke S, Meyer A, Huber CM, Witholt B, Wubbolts MG (1999) An alkane-responsive expression system for the production of fine chemicals. Appl Environ Microbiol 65:2324–2332

Panke S, Held M, Wubbolts MG, Witholt B, Schmid A (2002) Pilot-scale production of (*S*)-styrene oxide from styrene by recombinant *Escherichia coli* synthesizing styrene monooxygenase. Biotechnol Bioeng 80:33–41

Paul CE, Tischler D, Riedel A, Heine T, Itoh N, Hollmann F (2015) Nonenzymatic regeneration of styrene monooxygenase for catalysis. ACS Catal 5:2961–2965

Pedragosa-Moreau S, Archelas A, Furstoss R (1993) Microbial transformations. 28. Enantiocomplementary epoxide hydrolases as a preparative access to both enantiomers of styrene oxide. J Org Chem 58:5533–5536

Pedragosa-Moreau S, Archelas A, Furstoss R (1994) Microbiological transformations. 29. Enantioselective hydrolysis of epoxides using microorganisms: a mechanistic study. Bioorg Med Chem 2:609–616

Pedragosa-Moreau S, Archelas A, Furstoss R (1996a) Microbial transformations 32. Use of epoxide hydrolase mediated biohydrolysis as a way to enantiopure epoxides and vicinal diols: application to substituted styrene oxide derivatives. Tetrahedron 52:4593–4606

Pedragosa-Moreau S, Morisseau C, Zylber J, Archelas A, Baratti J, Furstoss R (1996b) Microbiological transformations. 33. Fungal epoxide hydrolases applied to the synthesis of enantiopure para-substituted styrene oxides. A mechanistic approach. J Org Chem 61:7402–7407

Qaed AA, Lin H, Tang D-F, Wu Z-L (2011) Rational design of styrene monooxygenase mutants with altered substrate preference. Biotechnol Lett 33:611–616

Qi WW, Vannelli T, Breinig S, Ben-Bassat A, Gatenby AA, Haynie SL, Sariaslani FS (2007) Functional expression of prokaryotic and eukaryotic genes in *Escherichia coli* for conversion of glucose to *p*-hydroxystyrene. Metab Eng 9:268–276

Renata H, Wang ZJ, Kitto RZ, Arnold FH (2014) P450-catalyzed asymmetric cyclopropanation of electron-deficient olefins under aerobic conditions. Catal Sci Technol 4:3640–3643

Riedel A, Heine T, Westphal AH, Conrad C, van Berkel WJH, Tischler D (2015) Catalytic and hydrodynamic properties of styrene monooxygenases from *Rhodococcus opacus* 1CP are modulated by cofactor binding. AMB Express 5:30

Rodriguez-Mozaz S, Lopez de Alda MJ, Barceló D (2006) Biosensors as useful tools for environmental analysis and monitoring. Anal Bioanal Chem 386:1025–1041

Rui L, Cao L, Chen W, Reardon KF, Wood TK (2004) Active site engineering of the epoxide hydrolase from *Agrobacterium radiobacter* AD1 to enhance aerobic mineralization of *cis*-1,2-dichloroethylene in cells expressing an evolved toluene *ortho*-monooxygenase. J Biol Chem 279:46810–46817

Rui L, Cao L, Chen W, Reardon KF, Wood TK (2005) Protein engineering of epoxide hydrolase from *Agrobacterium radiobacter* AD1 for enhanced activity and enantioselective production of (*R*)-1-phenylethane-1,2-diol. Appl Environ Microbiol 71:3995–4003

Ruinatscha R, Bühler K, Schmid A (2014) Development of a high performance electrochemical cofactor regeneration module and its application to the continuous reduction of FAD. J Mol Catal B Enzym 103:100–105

Santhanam L, Dordick JS (2002) Chloroperoxidase-catalyzed epoxidation of styrene in aqueous and nonaqueous media. Biocatal Biotransform 20:265–274

Schmid A, Dordick JS, Hauer B, Kiener A, Wubbolts M, Witholt B (2001) Industrial biocatalysis today and tomorrow. Nature 409:258–268

Spelberg JHL, Rink R, Kellogg RM, Janssen DB (1998) Enantioselectivity of a recombinant epoxide hydrolase from *Agrobacterium radiobacter*. Tetrahedron Asymmetry 9:459–466

Stiner L, Halverson LJ (2002) Development and characterization of a green fluorescent protein-based bacterial biosensor for bioavailable toluene and related compounds. Appl Environ Microbiol 68:1962–1971

Tieman DM, Loucas HM, Kim JY, Clark DG, Klee HJ (2007) Tomato phenylacetaldehyde reductases catalyze the last step in the synthesis of the aroma volatile 2-phenylethanol. Phytochemistry 68:2660–2669

Tischler D, Kaschabek SR (2012) Microbial degradation of xenobiotics. In: Singh SN (ed) Springer, Berlin, pp 67–99

Tischler D, Eulberg D, Lakner S, Kaschabek SR, van Berkel WJH, Schlömann M (2009) Identification of a novel self-sufficient styrene monooxygenase from *Rhodococcus opacus* 1CP. J Bacteriol 191:4996–5009

Tischler D, Kermer R, Gröning JAD, Kaschabek SR, van Berkel WJH, Schlömann M (2010) StyA1 and StyA2B from *Rhodococcus opacus* 1CP: a multifunctional styrene monooxygenase system. J Bacteriol 192:5220–5227

Toda H, Imae R, Itoh N (2012a) Efficient biocatalysis for the production of enantiopure (*S*)-epoxides using a styrene monooxygenase (SMO) and *Leifsonia* alcohol dehydrogenase (LSADH) system. Tetrahedron Asymmetry 23:1542–1549

Toda H, Imae R, Komio T, Itoh N (2012b) Expression and characterization of styrene monooxygenases of *Rhodococcus* sp. ST-5 and ST-10 for synthesizing enantiopure (*S*)-epoxides. Appl Microbiol Biotechnol 96:407–418

Toda H, Imae R, Itoh N (2014) Bioproduction of chiral epoxyalkanes using styrene monooxygenase from *Rhodococcus* sp. ST-10 (RhSMO). Adv Synth Catal 356:3443–3450

Toda H, Ohuchi T, Imae R, Itoh N (2015) Microbial production of aliphatic (*S*)-epoxyalkanes by using *Rhodococcus* sp. strain ST-10 styrene monooxygenase expressed in organic-solvent-tolerant *Kocuria rhizophila* DC2201. Appl Environ Microbiol 81:1919–1925

Utkin I, Yakimov M, Matveeva L, Kozlyak E, Rogozhin I, Solomon Z, Bez-borodov A (1991) Degradation of styrene and ethylbenzene by *Pseudomonas* species Y2. FEMS Microbiol Lett 77:237–242

van Berkel WJH, Kamerbeek NM, Fraaije MW (2006) Flavoprotein monooxygenases, a diverse class of oxidative biocatalysts. J Biotechnol 124:670–689

van Hellemond EW, Janssen DB, Fraaije MW (2007) Discovery of a novel styrene monooxygenase originating from the metagenome. Appl Environ Microbiol 73:5832–5839

van Loo B, Lutje Spelberg JH, Kingma J, Sonke T, Wubbolts MG, Janssen DB (2004) Directed evolution of epoxide hydrolase from *A. radiobacter* toward higher enantioselectivity by error-prone PCR and DNA shuffling. Chem Biol 11:981–990

Verhoef S, Wierckx N, Westerhof RGM, de Winde JH, Ruijssenaars HJ (2009) Bioproduction of *p*-hydroxystyrene from glucose by the solvent-tolerant bacterium *Pseudomonas putida* S12 in a two-phase water-decanol fermentation. Appl Environ Microbiol 75:931–936

Wang ZJ, Renata H, Peck NE, Farwell CC, Coelho PS, Arnold FH (2014) Improved cyclopropanation activity of histidine-ligated cytochrome P450 enables the enantioselective formal synthesis of levomilnacipran. Angew Chem Int Ed 53:6810–6813

Ward PG, de Roo G, O'Connor KE (2005) Accumulation of polyhydroxyalkanoate from styrene and phenylacetic acid by *Pseudomonas putida* CA-3. Appl Environ Microbiol 71:2046–2052

Ward PG, Goff M, Donner M, Kaminsky W, O'Connor KE (2006) A two step chemo-biotechnological conversion of polystyrene to a biodegradable thermoplastic. Environ Sci Technol 40:2433–2437

Warhurst AM, Clarke KF, Hill RA, Holt RA, Fewson CA (1994) Metabolism of styrene by *Rhodococcus rhodochrous* NCIMB 13259. Appl Environ Microbiol 60:1137–1145

Weijers CAGM (1997) Enantioselective hydrolysis of aryl, alicyclic and aliphatic epoxides by *Rhodotorula glutinis*. Tetrahedron Asymmetry 8:639–647

Wubbolts MG, Hoven J, Melgert B, Witholt B (1994a) Efficient production of optically-active styrene epoxides in 2-liquid phase cultures. Enzyme Microb Technol 16:887–894

Wubbolts MG, Reuvekamp P, Witholt B (1994b) Tol plasmid-specified xylene oxygenase is a wide substrate range monooxygenase capable of olefin epoxidation. Enzyme Microb Technol 16:608–615

Wuensch C, Gross J, Steinkellner G, Gruber K, Glueck SM, Faber K (2013) Asymmetric enzymatic hydration of hydroxystyrene derivatives. Angew Chem Int Ed 52:2293–2297

Wuensch C, Pavkov-Keller T, Steinkellner G, Gross J, Fuchs M, Hromic A, Lyskowski A, Fauland K, Gruber K, Glueck SM, Faber K (2015) Regioselective enzymatic β-carboxylation of *para*-hydroxystyrene derivatives catalyzed by phenolic acid decarboxylases. Adv Synth Catal 357:1909–1918

Zheng H, Reetz MT (2010) Manipulating the stereoselectivity of limonene epoxide hydrolase by directed evolution based on iterative saturation mutagenesis. J Am Chem Soc 132:15744–15751

Chapter 6
Conclusions and Future Perspectives

Abstract Since the first reports on microbial styrene degradation more and more interesting aspects were uncovered and led to the identification of highly adaptive microorganisms, so far unknown styrene catabolic routes, novel regulatory mechanisms, various enzymes involved in styrene biodegradation, and numerous biotechnological applications. All these aspects were described in the previous chapters, but, still there are some unsolved questions about degradation pathways, missing enzymes, or the evolution of regulatory and degradative elements. Furthermore, there are many ideas on possible applications of highlighted enzymes or bacteria. Some of these thoughts will be highlighted and discussed to initialize further research!

Keywords Styrene mineralization · Epoxidases and isomerases · Side-chain oxidation · Direct ring cleavage

6.1 Microbial Styrene Degradation and Its Open Questions

Two major aerobic routes for the bacterial styrene degradation have been reported and reviewed several times (Mooney et al. 2006; O'Leary et al. 2002b; Tischler and Kaschabek 2012; Warhurst and Fewson 1994). However, only the one via the vinyl side-chain attack and phenylacetic acid as central intermediate seems to have evolved specifically for styrene catabolism. It had been determined a few times in *Pseudomonas* species, and had been reported recently for a *Rhodococcus* and a *Sphingopyxis* strain (Oelschlägel et al. 2014b; Toda and Itoh 2012). This route does include the initial enzyme activities of a styrene monooxygenase (SMO), a styrene oxide isomerase (SOI), and a phenylacetaldehyde dehydrogenase (PAD),

© The Author(s) 2015
D. Tischler, *Microbial Styrene Degradation*,
SpringerBriefs in Microbiology, DOI 10.1007/978-3-319-24862-2_6

and results in the formation of phenylacetic acid as a central cell metabolite (Teufel et al. 2010). However, a recent study on the isolation of styrene-degrading bacteria harboring this route and along the key-enzymatic activity of an SOI (Oelschlägel et al. 2014b, c) revealed that only a minor portion of soil bacteria possess this side-chain pathway. The authors also screened for SOI-genes and confirmed the low abundance among bacteria investigated. Thus, it might be likely that most soil bacteria possess unspecific routes to degrade styrene, which need to be demonstrated, of course, on genetic as well as activity level in future studies. Further, the direct attack of styrene at the aromatic system was found earlier (Warhurst et al. 1994) and can be explained by the unspecific biotransformation of styrene via a typical *meta*-pathway (Patrauchan et al. 2008). Here it would be interesting to identify which *meta*-pathways are involved into styrene degradation, since so far only a biphenyl route had been investigated. Further, the possibility of an *ortho*-pathway cannot be ruled out, but evidence need to be provided. Thus the working hypothesis could be either a degradation of styrene via its vinyl side-chain or its aromatic nucleus via a *meta*-pathway might occur in nature.

Only for a few *Pseudomonas* species, namely *P. fluorescens* ST (Beltrametti et al. 1997; Marconi et al. 1996), *P. putida* CA-3 (O'Connor et al. 1995, 1997; O'Leary et al. 2001, 2002a), *P. taiwanensis* VLB120 (Otto et al. 2004; Panke et al. 1998; Vollmer et al. 2014) and *Pseudomonas* strains LQ26 (Lin et al. 2010, 2011b), S12 (Kantz et al. 2005; O'Connor et al. 1997), SN1 (Park et al. 2006), as well as Y2 (Utkin et al. 1991; Velasco et al. 1998) a detailed investigation of the styrene catabolic pathway including genetic as well as biochemical data have been reported. The more recent investigations of Toda and Itoh (2012) showed that in addition to pseudomonads also *Rhodococcus* sp. strain ST-5 harbors a complete *sty*-operon (*styABCD*) and evidence of activity from recombinant expressed proteins was provided. Previously, the wild-type strain had been investigated (Itoh et al. 1996). Interestingly, the wild-type SMO as well as the SOI from this strain are supposed to be cell wall bound (Itoh et al. 1997a; Toda and Itoh 2012; Toda et al. 2012). This observation is a somewhat outstanding feature since all other SMOs reported are soluble cytosolic proteins and the majority of SOIs are supposed to be membrane-linked (Mooney et al. 2006; Oelschlägel et al. 2012a, 2014b; Tischler and Kaschabek 2012). The corresponding proteins from strain ST-5 are more similar on amino acid level to their *Pseudomonas*-counterparts, and not as expected to the proteins from Actinobacteria reported earlier (Tischler et al. 2012). Further investigations are necessary to elucidate the abundance of the *sty*-gene cluster among other bacteria. In respect to our findings for the strain *Rhodococcus opacus* 1CP (Oelschlägel et al. 2012a, 2014b; Tischler et al. 2009), we assumed and later showed the presence of such a *sty*-operon, too (Riedel et al. 2015). From this gene cluster-derived genetic information allowed us to produce the recombinant SMO and SOI (Oelschlägel et al. 2015a; Riedel et al. 2015) and thus we have initially activity data determined for the SMO (StyA 80 mU mg^{-1} epoxidase activity; 75 StyB U mg^{-1} NADH:FAD oxidoreductase activity) and for the SOI (45 U mg^{-1}). Especially, for the SMO the values are higher compared to the activities of SMOs derived of other rhodococci (Toda et al. 2012) which makes the SMOs from strain 1CP candidates for more biocatalytic studies.

However, any data on the third enzyme of the upper styrene degradation route, the PAD, are still missing for strain 1CP, but, also for the most other strains investigated. Respectively, genome and transcriptome analysis from styrene-grown strain 1CP might provide further insights and reveal the presence of transcripts of these genes and so clarify their role in the active styrene degradation. Further, these studies might help to enlight if StyA1/StyA2B and/or StyA/StyB are involved actively in styrene activation for its metabolization. This question we raised already in our first publication on SMOs (Tischler et al. 2009) and so far the presence of StyA1/StyA2B is enigmatic and needs explanation.

Exemplary, a few related Actinobacteria comprise more than three different hypothetical SMO-genes (see phylogenetic analyses published earlier Tischler et al. 2009, 2012; van Hellemond et al. 2007). That fits well to the description of rhodococci as a rich reservoir ('catabolic powerhouse') for biocatalytic relevant pathways or genes which is caused by often reported gene redundancy of rhodococci (Gröning et al. 2014a; Larkin et al. 2005; McLeod et al. 2006; Warhurst and Fewson 1994). Also in these cases of multiple SMO-genes in bacteria, no functional verification is given so far. Thus their biocatalytic properties as well as metabolic role remains unclear.

The other route for styrene breakdown via a direct ring attack was first described in detail for *Rhodococcus rhodochrous* NCIMB 13259 (Warhurst et al. 1994). It is most likely a biphenyl degradation pathway (Patrauchan et al. 2008; Tischler and Kaschabek 2012) which gets induced by styrene and so employed to degrade the volatile compound. The belonging enzymes might have a rather broad substrate spectrum (Knackmuss et al. 1976; Marín et al. 2010; Warhurst et al. 1994) to facilitate these unspecific reactions. As it has been demonstrated there are many bacteria which likely use these unspecific *meta*-pathways (Oelschlägel et al. 2014b), but none is studied in detail to understand regulation and enzymatic activities. In addition, some fungi are able to grow on styrene as sole source of carbon. They use a slightly modified route via the vinyl site-chain as known for pseudomonads or the human detoxification pathway. Here only information on the initial styrene attack by means of a P450-styrene monooxygenase had been reported (Cox et al. 1996). The following enzymes involved in the pathway await their identification and description, respectively.

As already mentioned, only a small number of styrene-degrading microorganisms were characterized in detail (Oelschlägel et al. 2014b, c; Tischler and Kaschabek 2012), and hence only limited information are accessible on the distribution, regulation, and functionality of the styrene-specific degradation pathways as described for few pseudomonads for example. From an evolutionary point of view, it would be highly interesting to get a broader view on the biodegradability of styrene, the involved microorganisms, as well as the occurrence of additional styrene-specific pathways and their functionality. Furthermore, the observation that other pathways, like for biphenyl catabolism, allow also the activation and degradation of styrene demonstrates that rather unspecific routes might be often employed by microorganisms, too. Thus, several pathways might act in concert and allow the biodegradation of compounds like styrene.

In order to identify further microorganisms or enzymes with a considerable potential for biotechnological applications, dealing with styrene or its derivatives, several approaches seem to be suitable. Exemplary, the isolation of mixed or pure cultures from promising samples as contaminated soils and aquifers is favorable for the detection of highly (bio)active microorganisms for bioremediation. An overview about those studies in the field of biodegradation of volatile organic compounds (VOCs) with special emphasis on styrene as one of world wide most produced and processed chemical compounds was given elsewhere (Tischler and Kaschabek 2012). Recently, the isolation of a *Pseudomonas* strain designated as strain LQ26 by Lin et al. (2010) showed the potential of this conventional method, because this styrene-assimilating strain comprises a complete styrene-specific operon harboring a unique SMO with an unexpected substrate spectrum (Lin et al. 2010, 2011b). In addition, cultivation-independent approaches as (meta)genomics turn into focus and allow in combination with selective screening methods the detection of interesting and novel biocatalysts. By applying the technique, an interesting SMO was found and initially characterized (van Hellemond et al. 2007). Unfortunately, no complete gene cluster for the styrene degradation was identified from the cloned metagenomic fragments, showing a mentionable disadvantage of this method. A similar observation has been made in case of *Rhodococcus* sp. ST-10 for which also only SMO-genes had been determined (Toda and Itoh 2012). Often such genes which are located apart from a gene cluster cannot be assigned to a specific metabolic activity. Especially, in case of StyA1/StyA2B from strain 1CP the enzyme was classified according to its biocatalytic properties but a potential metabolic role could not be assigned (Tischler et al. 2009). Nowadays, the abundance of information accessible via a variety of databases allows beside the identification of single enzymes also the identification of complete pathways and may substitute or even replace above mentioned cultivation based approaches to determine novel and valuable biocatalysts. Thus a great reservoir of so far not characterized enzymes or even pathways gets accessible. Recently, such a database mining approach was reported for flavin-dependent proteins, but more from a general point of view as for biocatalysis (Macheroux et al. 2011). Interestingly, a huge number of oxidoreductases had been identified, whereas especially Actinobacteria like *Streptomyces* harbor many and show a distinct gene redundancy. This fits well for related rhodococci as mentioned above and exemplary *Rhodococcus jostii* RHA1 is also rich in oxygenases (203, predicted) (McLeod et al. 2006). However, these in silico-provided data have to be evaluated by biochemical experiments. The importance and potential of oxidative catalysts were recently presented in general reviews about oxidative enzymes (Arora et al. 2010; Behrens et al. 2011; Ceccoli et al. 2014; Hollmann et al. 2011; Huijbers et al. 2014; Lin et al. 2011a). These deal especially with the discovery, improvement, and application of such enzymes. In addition, the relevance of these biocatalysts in comparison to conventional chemical synthesis approaches was discussed.

Only little information are available on the anaerobic styrene degradation (see Sect. 2.3). Metabolites and thus catabolic routes of a limited number of microorganisms or consortia had been identified (Grbić-Galić et al. 1990; Araya et al. 2000). Further studies are necessary to first show the abundance of anaerobic styrene degradation among microorganisms. The presence of specially evolved electron chains and final electron acceptors need to be identified. Then of course genetic and biochemical studies are mandatory to enlighten the regulatory machinery and the mechanistic background. Here, even it is yet impossible to say or predict: there are styrene-specific pathways present or unspecific routes used to degrade styrene in absence of oxygen?

The degradation of styrene oligomers or even styrene polymers has been neither under aerobic nor anaerobic conditions extensively studied. A short summary was provided in Sect. 2.4. Further studies would be needed to clearly show how these compounds can be attacked and degraded by microorganisms. However, it seems obvious from results presented, that none of these routes discussed for the styrene monomer degradation are employed to catabolise oligomers of styrene. And only a single study indicates that styrene polymer can be degraded (Mor and Sivan 2008).

6.2 Application of Whole-Cells in Biocatalysis

A number of applications derived from either styrene-assimilating microorganisms themselves or enriched enzymes of corresponding pathways highlights the potential of styrene degraders (Fig. 6.1).

Microbial consortia can be employed in different bioremediation processes for the degradation (or detoxification) of styrene or mixtures of VOCs (summarized previously by Tischler and Kaschabek 2012). Besides the scientific studies on these topics, which deal with the optimization of reactors as wells as the operational parameters and the enrichment of highly active biomass, such approaches are already in industry established and run effectively. Companies as the Wilhelm Schimmel Pianofortefabrik GmbH (Braunschweig, Germany) have to deal with mixtures of VOCs during product manufacture. This company is mentioned because they have an environmental friendly program and are already running a biological air filter (personal communication, Mr. Zipp; Fig. 6.1). The natural presence of styrene and the anthropogenic release are likely reasons for the ubiquitous presence of styrene-catabolic activities in the environment (see Chaps. 1 and 2). Thus styrene was herein classified as a readily biodegradable compound. Companies as mentioned above usually make use of the natural occurring microbial flora and biodiversity for the inoculation of a biofilter for treating wastes (gas). So a microbial consortium is naturally present and will develop towards a powerful degrading machinery. Thus a further development is often not necessary, but, from an ecological point of view it would be interesting to identify microbial key-players in dependence of the waste streams and other conditions. Usually, the

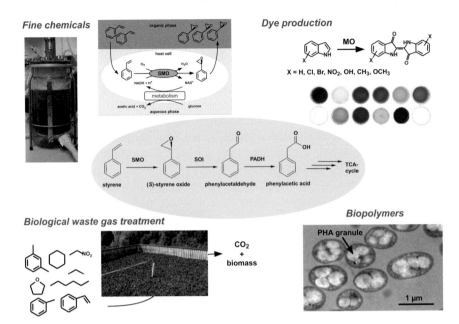

Fig. 6.1 Overview on applications by means of styrene-degrading microorganisms or their enzymes. Exemplary are shown (i) the whole-cell biotransformation with an SMO in a two-phase system for (S)-styrene oxide production (adapted from Kuhn et al. 2010), (ii) the conversion of indole-derivatives by a flavin-containing monooxygenase (MO) to indigoid dyes (adapted from Rioz-Martínez et al. 2011), which is also achievable applying SMOs, (iii) the topview on a biological waste gas filter plant from the company Wilhelm Schimmel Pianofortefabrik GmbH (Braunschweig, Germany) (picture adapted from Schröter 2006), and (iv) the formation of poly-hydroxyalkanoates (PHAs) by a styrene-degrading pseudomonad, which accumulate as granules in the cells and are promising biopolymers (adapted from Ward et al. 2005)

application of optimized strains within special filter-system is restricted to more persistent compounds or mixtures, not discussed herein.

Another approach can be the use of styrene-assimilating bacteria to produce valuable products while growing on styrene or derivatives as sole source of carbon and energy. Unbalanced nutrient supply such as the excess of a carbon source stimulates many bacteria to produce polyhydroxyalkanoates (PHAs) as storage compounds (Anderson and Dawes 1990; Madison and Huisman 1999). This was also observed for a pseudomonad comprising a functional *sty*-operon and allowed the production of such biodegradable polymers from styrene or even from thermally pretreated polystyrene (Ward et al. 2005, 2006). However, the low yield of PHAs (23 % of cell dry weight) from styrene-grown biomass seems so far not competitive enough for a biotechnological application and further optimization is necessary (Anderson and Dawes 1990; Ward et al. 2005). Besides the production of biopolymers via the styrene catabolism of soil bacteria one could also think about the co-metabolic production of valuable compounds. Styrene degrading bacteria can be fed with substrates, which are converted just by a single or a set of enzyme(s) from the pathway and the formed products will accumulate during the fermentation process. For instance, indole or derivatives can be applied as co-substrates and indigoid dyes will accumulate in the media (O'Connor et al. 1997). Also some substituted styrene derivatives may so provide access to corresponding phenylacetic acids as recently demonstrated by wild-type and recombinant strains (Oelschlägel et al. 2014c, 2015a, b). This strategy can be further developed to produce valuable compounds as ibuprofen for example.

The ability of some *Pseudomonas* species to form rather stable biofilms in combination with genetic engineering can yield efficient whole-cell biocatalysts for the production of fine chemicals as styrene oxide, whereas the microorganisms in a biofilm seem to be more resistant against toxic compounds (Gross et al. 2007, 2010; Halan et al. 2010, 2011). The reachable volume productivity with such a system is of course lower as from stirred-tank reactors. But the biofilm approach allows a more stable and continuous performance, which provides economic and environmental advantages. The switch from batch or fed-batch to more continuously running systems was intensively discussed among the biotechnology community (1st European Congress of Applied Biotechnology, 2011) and is supposed to be the next step to a more efficient production in the field. For fine chemical syntheses and smaller scales, biofilm-based systems present a promising alternative. However, the applicability of biofilms for industrial scale production remains to be answered.

A number of enzymes from styrene degrading microorganisms and corresponding pathways can be applied to biocatalysis in order to produce valuable products (reviewed by Mooney et al. 2006; O'Leary et al. 2002b; Tischler and Kaschabek 2012; see Chaps. 4 and 5). Enzymes as SMO, SOI, epoxide hydroxylases, phenylacetaldehyde reductase, or extra-diol-acting dioxygenases allow the efficient and regioselective formation of epoxides, sulfoxides, aldehydes, alcohols, diols, catechols, and indigoid dyes. Depending on the source of the enzymes and applied substrates, the conversions can be enantioselective, which is important especially for agrochemical, pharmaceutical, as well as food industry, where some of these building blocks are of importance (Archelas and Furstoss 1997; Badone and Guzzi 1994; Besse and Veschambre 1994; Breuer et al. 2004; Hattori et al. 1995; Hollmann et al. 2011; Li et al. 2002; Lin et al. 2011a; Rao et al. 1992; Schmid et al. 2001; Torres Pazmiño et al. 2010).

The progress in several fields as identifying novel biocatalysts (by in silico, in vitro, or in vivo techniques), cofactor regeneration (enzymatic and non-enzymatic), enzyme immobilization (by cross-linking enzymes, coating enzymes on porous material), directed evolution in combination with novel screening systems, and the resourcing of solvent tolerant strains for catalysis promises a bright future for biotechnology as a tool within the field of synthetic industrial chemistry (Arora et al. 2010; Breuer et al. 2004; Lin et al. 2011a; Hollmann et al. 2011; Schmid et al. 2001; Schulze and Wubbolts 1999; Sheldon 2007; Torres Pazmiño et al. 2010).

6.3 Styrene Monooxygenases and Future Expectations

The SMOs are the best studied enzymes obtained from the styrene degradation pathway. Furthermore, there are more related enzymes described which are likely not involved into the styrene degradation as StyA1/StyA2B. For these enzymes general mechanistic details and structural insights have been reported and a large biotechnological campaign describes all aspects how to make use of these enzymes. However, there are still some open questions regarding SMOs and those are now highlighted.

6.3.1 Sulfoxidation as the Promising Application

Various classes of enzymes perform sulfoxidation reactions in an enantioselective manner as exemplary drawn in Fig. 6.2a for the prochiral methylphenylsulfide (thioanisole). Valuable products can be produced by the application of enzymes like peroxidases and other heme-proteins (Dai and Klibanov 2000; Tuynman et al. 1998, 2000), P450-dependent monooxygenases (Zhang et al. 2010), Baeyer-Villiger monooxygenases (Rioz-Martínez et al. 2010), flavin-containing monooxygenases (FMOs) (Rioz-Martínez et al. 2011), as well as tyrosinase (Pievo et al. 2008) on prochiral sulfides (Fig. 6.2b). Depending on the applied enzyme class cofactors or even further substrates are necessary and reaction mechanisms may vary.

Earlier studies demonstrated that SMOs perform besides epoxidation also sulfoxidation reactions (summarized by Montersino et al. 2011). A few alkyl aryl

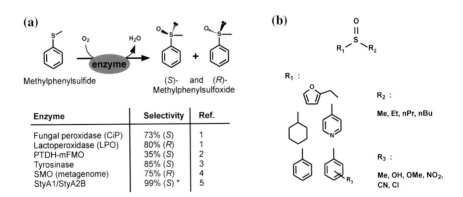

(a)

Methylphenylsulfide   (S)- and (R)-
                                       Methylphenylsulfoxide

Enzyme	Selectivity	Ref.
Fungal peroxidase (CiP)	73% (S)	1
Lactoperoxidase (LPO)	80% (R)	1
PTDH-mFMO	35% (S)	2
Tyrosinase	85% (S)	3
SMO (metagenome)	75% (R)	4
StyA1/StyA2B	99% (S) *	5

(b)

R_1 :

R_2 :

Me, Et, nPr, nBu

R_3 :

Me, OH, OMe, NO_2, CN, Cl

Fig. 6.2 Enzyme-based sulfoxidations. **a** The oxidation of thioanisole by various biocatalysts yielding different degrees of (enantio)selectivity is shown as enantiomeric excess (*1*: Tuynman et al. 1998; *2*: Rioz-Martínez et al. 2011; *3*: Pievo et al. 2008; *4*: van Hellemond et al. 2007; *5*: Paul et al. 2015). *This high ee-value represents the best result reported for a monooxygenase catalyzed sulfoxidation so far. **b** Valuable products accessible by enzymatic sulfoxidations are highlighted (Rioz-Martínez et al. 2010, 2011; van Hellemond et al. 2007)

sulfides were converted almost with a similar rate during catalysis as styrene or even faster. However, the enantioselectivity of methylphenylsulfide oxidation by StyA from *Pseudomonas taiwanensis* strain VLB120 (up to 26 % e.e., the major enantiomer was not determined) (Hollmann et al. 2003) as well as that towards ethylphenylsulfide oxidation by a SMO from metagenome (up to 92 % e.e., (*R*)-form) (van Hellemond et al. 2007) reached in both cases not that enantioselectivity of styrene epoxidation of >99 % e.e., in which the (*S*)-epoxide was always formed. In cooperation with the group of Frank Hollmann (TU Delft, The Netherlands) we found in preliminary studies that the SMO StyA1/StyA2B from *R. opacus* 1CP is able to convert methylphenylsulfide into the (*S*)-form of the corresponding sulfoxide with an enantiomeric excess (e.e.) >99 % (Fig. 6.2a). Further studies confirmed the result obtained (Paul et al. 2015) and provide crystallization points for future investigations. Even mechanistic and structural studies of this sulfoxidase activity might be of interest in order to determine the relevant amino acids causing the high (enantio)selectivity. And of course, the number of substrates tested needs to be increased to provide access towards industrial relevant products.

6.3.2 Structural and Mechanistic Questions

So far only little is known about the structure of SMOs (Huijbers et al. 2014). The analysis of available sequences indicated an evolutionary linkage to class A of flavoprotein monooxygenases (represented by the prototype 4-OH-benzoate hydroxylase: PHBH), since both classes encode one specific dinucleotide binding domain (Rossmann fold) (Montersino et al. 2011; van Berkel et al. 2006). First modeling and later a published X-ray structure provided some insights (Feenstra et al. 2006; Ukaegbu et al. 2010) and confirmed that SMOA (epoxidase component of a SMO of type E1) has in general the same fold as PHBH. Unfortunately, the SMOA structure was obtained without bound substrate (styrene) or cofactor (FAD). Therefore, the exact location of substrate and cofactor within the active site can only be simulated and assumed by molecular docking experiments (Feenstra et al. 2006; Qaed et al. 2011). Further structural studies are needed in order to highlight the active site and crucial amino acid residues for substrate binding as well as for catalysis. Recent mechanistic studies regarding a SMO (type E1) from a *Pseudomonas* strain revealed a number of interesting findings (Kantz and Gassner, 2011; Morrison et al. 2013). In contrast to PHBH and related enzymes, the SMO epoxidases do not bind or interact tightly with oxidized FAD. Instead, they have solely a high affinity for the reduced cofactor. The strictly NADH-dependent FAD reductase SMOB generates rapidly reduced FAD which is supposed to be transferred to SMOA by diffusion or by a mixed mechanism of diffusion and direct transfer (Kantz et al. 2005; Morrison et al. 2013). From these studies also, a structure of SMOB is now available and allows at least seeing the FAD binding and likely SMOB–SMOB interaction. Here, the NADH-binding mode would be of interest as well. Furthermore, the interaction of SMOA and SMOB had been several times suggested and some experimental evidence was provided, but a structure of the complex or any determination of interaction by means of labeling or other approaches is still missing.

First mechanistic insights for the other type of SMOs (E2) were gained from herein presented studies (Tischler et al. 2010, 2011, 2013). The monooxygenase components StyA1 as well as StyA2B perform the epoxidation of styrene by means of reduced FAD, which allows the activation of molecular oxygen. From first spectroscopic as well as kinetic data in combination with numeric modeling approaches, we suggested that the reaction sequence with FAD-intermediates is similar to that observed for SMOA from pseudomonads (Kantz and Gassner 2011; Tischler et al. 2011, 2013). In addition, we evaluated that StyA1 performs in presteady state as well as in steady-state reactions, a much faster epoxidation than StyA2B. An unproductive reoxidation of a peroxyflavin intermediate was observed for both epoxidases. Thus likely hydrogen peroxide was formed during these presteady state experiments, which is the so-called uncoupling. However, the low styrene oxygenating activity of StyA2B and therewith the role of the StyA2-component within this monooxygenase is still unclear. A more regulatory or even stabilizing function during the catalysis was suggested, but need to be demonstrated. Unfortunately, no structural data are so far available for E2-type SMOs.

6.3.3 Evolution of Styrene Monooxygenases

The lack of detailed structural information especially on the substrate binding mode hampers at the moment the directed evolution of SMOs by rational design. However, the nowadays huge computational power and innovative programs allow generating and screening virtual mutant libraries for improved properties (Behrens et al. 2011) or even alternative activities (Steinkellner et al. 2014). Thus applying molecular dynamic methods for in silico docking, the substrate and/or cofactor into the active site of a protein can provide substantial information. Such an approach for SMOA allowed Qaed et al. (2011) to determine amino acids which are probably relevant for substrate recognition and conversion. Indeed, they were able to alter the substrate specificity of SMOA toward the more bulkier α-ethylstyrene via site-directed mutagenesis (in respect to styrene).

Besides the mentioned site-directed approaches to change the specificity and/or activity of SMOs, another technique for in vitro evolution has been conducted by Gursky et al. (2010). They subjected the SMO-genes from *Pseudomonas putida* CA-3 to three rounds of error-prone polymerase chain reaction in order to improve the specific activity. The yielded clones were prescreened via their capability to convert indole which leads to the formation of the dark-blue dye indigo (O'Connor et al. 1997). Clones which produced rapidly indigo were also assayed for their activity toward styrene and indene, since the produced epoxides (*S*)-styrene oxide and (1*S*,2*R*)-indene oxide are valuable building blocks for chemical industries (Tischler and Kaschabek 2012). The generated variants comprising the highest epoxidation activity showed indeed an 8-fold and a 12-fold increased activity for the substrates styrene and indene, respectively. The genes of these improved biocatalysts were analyzed for mutations in respect to changes close to or within the active site. For comparison of mutants to the ancestral SMO served, a model structure (PDB-file: 2HD8; Feenstra et al. 2006) and possible effects of substituted amino acids were discussed. But as mentioned above due to the lack of detailed information of the active site including bound substrate or cofactor these suggestions have to be taken with care. Later studies might reinforce these assumptions and allow performing a more directed evolution.

More recently, Lin et al. (2012) changed several residues in SMOA based on the StyA-structure (pdb: 3IHM) and substrate docking approaches to alter substrate specificity and selectivity. This study provides first evidence on amino acids relevant for substrate binding and stereoselective conversion. Respectively, more such experimental approaches to alter the SMOA sequence in order to identify crucial residues for catalysis will help to draw the active site and bring it in line with the mechanistical studies reported earlier.

6.3.4 StyA1 and StyA2B—Origin and Physiological Role

As in detail described in Sect. 4.1, StyA1 and StyA2B represent the prototype of a second group of SMOs. The major question we still not solved is on the origin of this SMO and its physiological role in the organisms (Tischler et al. 2009, 2012). First, the fusion character of StyA2B was supposed to be the characteristic feature to discriminate these enzymes, but later we could demonstrate also systems comprising a single epoxidase and reductase belong to type E2 SMOs (Gröning et al. 2014b; Tischler et al. 2012). The phylogenetic studies indicate SMOs may have evolved from a single predecessor, but, into two different directions E1 and E2 type SMOs. And the fusion character of StyA2B-like proteins might have evolved several times as well and seems not only a product of horizontal gene transfer. However, here many more studies are needed to first understand the evolution but also to determine the physiological role of these proteins.

6.4 Styrene Oxide Isomerase in Biocatalysis

So far, only a few representative enzymes for the class of the SOIs have been described in the literature (Hartmans et al. 1989; Itoh et al. 1997a; Liao 2011; Miyamoto et al. 2007; Oelschlägel et al. 2012a, b, 2014b, c, 2015a; Toda and Itoh 2012). Especially, the access to suitable amounts of the enzyme from either recombinant or wild-type hosts seemed to be a limiting factor for the characterization or an application (Oelschlägel et al. 2012a, b). However, they all share features making them interesting for biocatalysis: (i) high turnover numbers, (ii) process stability, (iii) enantioselectivity, and (iv) independence of cofactors.

SOIs are rather small proteins of about 18 kDa and perform the cofactor-independent intramolecular isomerization (Meinwald-rearrangement like reaction; Meinwald et al. 1963) of styrene oxide into phenylacetaldehyde (Hartmans et al. 1989; Miyamoto et al. 2007). A high specific activity (up to 300 U mg^{-1}; which can be tuned further by thermal activation; Oelschlägel et al. 2012a) and the declared stability of these enzymes made them interesting for biotechnologists as mentioned above. The stability and applicability had been further improved by immobilization on porous carriers (Oelschlägel et al. 2014a). This is one of the first studies of a covalent linkage of a membrane protein to such carriers. Therefore, the actual binding mode and three-dimensional shape of the product is of interest, but remains open for future studies. The independence of cofactors is another major advantage for a biocatalyst and allows operation under many conditions. However, during purification either from wild-type or recombinant host a red protein fraction had been obtained (M. Oelschlägel and S. Liao, personal communication). It is supposed that the red fraction is a heme-dependent protein of the respiratory chain which is also membrane linked and gets co-purified with the SOI protein. Respectively, the nature needs to be determined by further experiments. The other enzymes of the upper styrene pathway are dependent on NAD$^+$ or NADH/FAD, respectively, and here adjustments not only to the enzyme but also to the cofactor are necessary. It is also beneficial that only the formation of the product phenylacetaldehyde is observed and not the reverse reaction by SOI activity. In addition, further styrene oxide derivatives can be converted into the corresponding phenylacetaldehyde analogous compounds or even to prochiral ketones (Miyamoto et al. 2007; Oelschlägel et al. 2014c). The products obtained from these biotransformations are of high purity and valuable building blocks or flavor-like compounds as phenylacetaldehyde itself. The enzyme is supposed to be membrane bound and less expressed either from wild-type or recombinant hosts (Itoh et al. 1997a; Liao 2011; Nikodinovic-Runic et al. 2009). So far, a cell-free expression has not been demonstrated and a eukaryotic system was tested only once without success (Oelschlägel et al. 2015a). Thus further studies on the expression and production of this biocatalyst would be needed to increase the amount accessible.

Our recent investigations provide now a simple as well as effective protocol for the enrichment of even less expressed SOIs from different microorganisms (Oelschlägel et al. 2012b, 2014b). However, a complete purification of the protein from *R. opacus* 1CP, *P. fluorescens* ST or *S. fribergensis* Kp5.2 was not achieved, which is likely due to a highly hydrophobic character of the SOI-proteins. This hydrophobic nature as well as the membrane association of the SOI should not be seen as a major disadvantage since it allows to highly enrich the enzyme during purification approaches as well as from enzyme assays. The latter finding provides the possibility to reuse the enzyme preparation in several biotransformations demonstrated with an immobilized variant as well (Oelschlägel et al. 2014a). It would be interesting to repeat such studies with higher quantities of the enzyme to see on pilot scale how much product per batch process can be gained and how often the biocatalyst might be recycled. Furthermore, we could demonstrate the high activity as well as stability of the enzymes. Interestingly, a high substrate concentration caused no significant effects on the enzyme activity, whereas the product can yield an irreversible inhibition. The high reactivity of the phenylacetaldehyde and so covalently enzyme modifications might be the reason of the observed inhibitory effect. Here, mass spectrometry in combination with limited proteolysis might support to identify the center of inactivation. If possible, these amino acids might be changed by mutagenesis to create a biocatalyst which is not inactivated by its own product.

Especially, the SOI obtained from *Xanthobacter* sp. 124X need to be mentioned (Hartmans et al. 1989). As reported earlier, it was supposed to be a cell wall and not membrane linked as the other SOI-proteins. In a recent study, the postulated localization was confirmed (Oelschlägel et al. 2014b) and furthermore the corresponding gene and/or gene cluster must have drastical differences on sequence level to all the other bacteria since it was impossible to localize a respective SOI-gene from *Xanthobacter* sp. 124X.

However, with the now established protocol and therewith access to applicable amounts of protein more representatives can be biochemically characterized and this can provide further insights for this class of enzymes. Especially, the inhibitory effect of phenylacetaldehyde needs to be investigated.

6.5 Enzymes with an Enigmatic Background

As mentioned in Chaps. 2 and 3 the styrene-specific pathway yielding phenylacetic acid is well studied and enzymes have been described to a large extent. However, some variants of this pathway employ enzymes which are underrepresented and thus only little is known about their biochemistry, mechanisms and structural aspects.

Instead of a conventional SMO also a cytochrome P450-dependent monooxygenase had been identified to convert styrene to styrene oxide in the host *Exophilia jeanselmei* (Cox et al. 1996). It also needs FAD and oxygen to allow styrene epoxidation but utilizes only NADPH as source of electrons. The latter is a major difference to the conventional SMO as StyA/StyB which accepts solely NADH. In addition, the oxygen activation is here supposed to occur via the heme-component and not via the FAD itself which just delivers electrons stepwise to the heme. However, no more information on the abundance and behavior of this enzyme has been reported. It is not clear, if it is a styrene-specific P450-system or a more general or unspecific monooxygenase. The detailed mechanism of electron transfer, interaction with a reductase or other proteins as well as the epoxidation cycle has not been studied yet. Thus more studies are needed to demonstrate if it is an exception or a general route among fungi.

Two more enzymes act on intermediates of the pathway and had been designated as styrene oxide reductase (SOR) and phenylacetaldehyde reductase (PAR) (Tischler and Kaschabek 2012). The SOR-activity had been described once (Marcon et al. 1996) and has only been determined under certain incubation conditions. It yields 2-phenylethanol and was suggested presenting an alternative route in the styrene catabolism of *Pseudomonas fluorescens* ST. Here a genomic fragment was found to encode for the respective activity while the gene was not assigned. To clarify its background, the gene and its natural expression should be studied. Thus it can be studied in respect to styrene degradation as well. Of course, any studies on the protein itself would support its classification. In case of PAR, the story is very similar, but, here more data are already available since its presence in the wild-type microorganism (Itoh et al. 1997b) was confirmed by activity measurements and molecular biological investigations (Wang et al. 1999a, b). However, a true role in styrene catabolism needs to be verified, for example, by transcriptomics or gene-knock-out studies. Activity data clearly demonstrate its activity on phenylacetaldehyde, but the reverse reaction could not be determined. The latter is somewhat strange since the enzyme is per definition an alcohol dehydrogenase and should act also on 2-phenylethanol, respectively. Here, more sophisticated analyses are needed to completely understand its mechanism.

Epoxide hydrolases have been found to play a role in styrene detoxification and degradation by fungi (Braun-Lüllemann et al. 1997). The general mechanism and biotechnological applicability is well known (see Sect. 4.5). But, their role in styrene mineralization still is enigmatic since only for a few fungi such an activity had been determined. Induction pattern and interaction with the other degradative enzymes has not been studied so far. Thus more investigations regarding their role in degradation and their regulation need to be conducted.

References

Anderson AJ, Dawes EA (1990) Occurrence, metabolism, metabolic role, and industrial use of bacterial polyhydroxyalkanoates. Microbiol Rev 54:450–472

Araya A, Chamy R, Mota M, Alves M (2000) Biodegradability and toxicity of styrene in the anaerobic digestion process. Biotechnol Lett 22:1477–1481

Archelas A, Furstoss R (1997) Synthesis of enantiopure epoxides through biocatalytic approaches. Annu Rev Microbiol 51:491–525

Arora PK, Srivastava A, Singh VP (2010) Application of monooxygenases in dehalogenation, desulphurization, denitrification and hydroxylation of aromatic compounds. J Bioremed Biodegrad 1:112–119

Badone D, Guzzi U (1994) Synthesis of the potent and selective atypical beta-adrenergic agonist SR 59062 A. Bioorg Med Chem Lett 16:1921–1924

Behrens GA, Hummel A, Padhi SK, Schätzle S, Bornscheuer UT (2011) Discovery and protein engineering of biocatalysts for organic synthesis. Adv Synth Catal 353:2191–2215

Beltrametti F, Marconi AM, Bestetti G, Galli E, Ruzzi M, Zennaro E (1997) Sequencing and functional analysis of styrene catabolism genes from *Pseudomonas fluorescens* ST. Appl Environ Microbiol 63:2232–2239

Besse P, Veschambre H (1994) Chemical and biological synthesis of chiral epoxides. Tetrahedron 50:8885–8927

Braun-Lüllemann A, Majcherczyk A, Huttermann A (1997) Degradation of styrene by white-rot fungi. Appl Microbiol Biotechnol 47:150–155

Breuer M, Ditrich K, Habicher T, Hauer B, Kesseler M, Stuermer R, Zelinski T (2004) Industrial methods for the production of optically active intermediates. Angew Chem Int Ed 43:788–824

Ceccoli RD, Bianchi DA, Rial DV (2014) Flavoprotein monooxygenases for oxidative biocatalysis: recombinant expression in microbial hosts and applications. Frontiers Microbiology 5:1–14

Cox HHJ, Faber BW, van Heiningen WNM, Radhoe H, Doddema HJ, Harder W (1996) Styrene metabolism in *Exophiala jeanselmei* and involvement of a cytochrome P-450-dependent styrene monooxygenase. Appl Environ Microbiol 62:1471–1474

Dai L, Klibanov AM (2000) Peroxidase-catalyzed asymmetric sulfoxidation in organic solvents versus in water. Biotechnol Bioeng 70:353–357

Feenstra KA, Hofstetter K, Bosch R, Schmid A, Commandeur JNM, Vermeulen NPE (2006) Enantioselective substrate binding in a monooxygenase protein model by molecular dynamics and docking. Biophys J 91:3206–3216

Grbić-Galić D, Churchman-Eisel N, Mraković I (1990) Microbial transformation of styrene by anaerobic consortia. J Appl Bacteriol 69:247–260

Gröning JAD, Eulberg D, Tischler D, Kaschabek SR, Schlömann M (2014a) Gene redundancy of two-component (chloro)phenol hydroxylases in *Rhodococcus opacus* 1CP. FEMS Microbiol Lett 361:68–75

Gröning JAD, Kaschabek SR, Schlömann M, Tischler D (2014b) A mechanistic study on SMOB-ADP1: an NADH:flavin oxidoreductase of the two-component styrene monooxygenase of *Acinetobacter baylyi* ADP1. Arch Microbiol 196:829–845

Gross R, Hauer B, Otto K, Schmid A (2007) Microbial biofilms: new catalysts for maximizing productivity of long-term biotransformations. Biotechnol Bioeng 98:1123–1134

Gross R, Lang K, Bühler K, Schmid A (2010) Characterization of a biofilm membrane reactor and its prospects for fine chemical synthesis. Biotechnol Bioeng 105:705–717

Gursky LJ, Nikodinovic-Runic J, Feenstra KA, O'Connor KE (2010) In vitro evolution of styrene monooxygenase from *Pseudomonas putida* CA-3 for improved epoxide synthesis. Appl Microbiol Biotechnol 85:995–1004

Halan B, Schmid A, Bühler K (2010) Maximizing the productivity of catalytic biofilms on solid supports in membrane aerated reactors. Biotechnol Bioeng 106:516–527

Halan B, Schmid A, Bühler K (2011) Real-time solvent tolerance analysis of *Pseudomonas* sp. strain VLB120ΔC catalytic biofilms. Appl Environ Microbiol 77:1563–1571

Hartmans S, Smits JP, van der Werf MJ, Volkering F, de Bont JAM (1989) Metabolism of styrene oxide and 2-phenylethanol in the styrene-degrading *Xanthobacter* strain 124X. Microbiology (UK) 55:2850–2855

Hattori K, Nagano M, Kato T, Nakanishi I, Imai K, Kinoshita T, Sakane K (1995) Asymmetric synthesis of FR165914: a novel beta-3-adrenergic agonist with a benzocycloheptene structure. Bioorg Med Chem Lett 5:2821–2824

Hollmann F, Lin P-C, Witholt B, Schmid A (2003) Stereospecific biocatalytic epoxidation: the first example of direct regeneration of a FAD-dependent monooxygenase for catalysis. J Am Chem Soc 125:8209–8217

Hollmann F, Arends IWCE, Bühler K, Schallmey A, Bühler B (2011) Enzyme-mediated oxidations for the chemist. Green Chem 13:226–265

Huijbers MME, Montersino S, Westphal AH, Tischler D, van Berkel WJH (2014) Flavin dependent monooxygenases. Arch Biochem Biophys 544:2–17

Itoh N, Yoshida K, Okada K (1996) Isolation and identification of styrene-degrading *Corynebacterium* strains, and their styrene metabolism. Biosci Biotechnol Biochem 60:1826–1830

Itoh N, Hayashi K, Okada K, Ito T, Mizuguchi N (1997a) Characterization of styrene oxide isomerase, a key enzyme of styrene and styrene oxide metabolism in *Corynebacterium* sp. Biosci Biotechnol Biochem 61:2058–2062

Itoh N, Morihama R, Wang J, Okada K, Mizuguchi N (1997b) Purification and characterization of phenylacetaldehyde reductase from a styrene-assimilating *Corynebacterium* strain, ST-10. Appl Environ Microbiol 63:3783–3788

Kantz A, Gassner GT (2011) Nature of the reaction intermediates in the flavin adenine dinucleotide-dependent epoxidation mechanism of styrene monooxygenase. Biochemistry 50:523–532

Kantz A, Chin F, Nallamothu N, Nguyen T, Gassner GT (2005) Mechanism of flavin transfer and oxygen activation by the two-component flavoenzyme styrene monooxygenase. Arch Biochem Biophys 442:102–116

Knackmuss H-J, Hellwig M, Lackner H, Otting W (1976) Cometabolism of 3-methylbenzoate and methylcatechols by a 3-chlorobenzoate utilizing *Pseudomonas*: accumulation of (+)-2,5-dihydro-4-methyl- and (+)-2,5- dihydro-2-methyl-5-oxo-furan-2-acetic acid. Eur J Appl Microbiol 2:267–276

Kuhn D, Kholiq MA, Heinzle E, Bühler B, Schmid A (2010) Intensification and economic and ecological assessment of a biocatalytic oxyfunctionalization process. Green Chem 12:815–827

Larkin MJ, Kulakov LA, Allen CCR (2005) Biodegradation and *Rhodococcus*—masters of catabolic versatility. Curr Opin Biotechnol 16:282–290

Li Z, van Beilen JB, Duetz WA, Schmid A, de Raadt A, Griengl H, Witholt B (2002) Oxidative biotransformations using oxygenases. Curr Opin Chem Biol 6:136–144

Liao Chan SA (2011) Purification and characterization of recombinant styrene oxide isomerase from *Pseudomonas p.* 12. Master Thesis, San Francisco State University, California

Lin H, Qiao J, Liu Y, Wu Z-L (2010) Styrene monooxygenase from *Pseudomonas* sp. LQ26 catalyzes the asymmetric epoxidation of both conjugated and unconjugated alkenes. J Mol Catal B Enzym 67:236–241

Lin H, Liu J-Y, Wang H-B, Qaed Ahmed AA, Wu ZL (2011a) Biocatalysis as an alternative for the production of chiral epoxides: a comparative review. J Mol Catal B Enzym 72:77–89

Lin H, Liu Y, Wu ZL (2011b) Highly diastereo- and enantio-selective epoxidation of secondary allylic alcohols catalyzed by styrene monooxygenase. Chem Commun (Camb) 47:2610–2612

Lin H, Tang D-F, Qaed Ahmed AA, Liu Y, Wu Z-L (2012) Mutations at the putative active cavity of styrene monooxygenase: enhanced activity and reversed enantioselectivity. J Biotechnol 161:235–241

Macheroux P, Kappes B, Ealick SE (2011) Flavogenomics—a genomic and structural view of flavin-dependent proteins. FEBS J 278:2625–2634

Madison LL, Huisman GW (1999) Metabolic engineering of poly (3-hydroxyalkanoates): from DNA to plastic. Microbiol Mol Biol Rev 63:21–53

Marconi AM, Beltrametti F, Bestetti G, Solinas F, Ruzzi M, Galli E, Zennaro E (1996) Cloning and characterization of styrene catabolism genes from *Pseudomonas fluorescens* ST. Appl Environ Microbiol 62:121–127

Marín M, Pérez-Pantoja D, Donoso R, Wray V, González B, Pieper DH (2010) Modified 3-oxoadipate pathway for the biodegradation of methylaromatics in *Pseudomonas reinekei* MT1. J Bacteriol 192:1543–1552

McLeod MP, Warren RL, Hsiao WWL, Araki N, Myhre M, Fernandes C, Miyazawa D, Wong W, Lillquist AL, Wang D, Dosanjh M, Hara H, Petrescu A, Morin RD, Yang G, Stott JM, Schein JE, Shin H, Smailus D, Siddiqui AS, Marra MA, Jones SJM, Holt R, Brinkman FSL, Miyauchi M, Fukuda M, Davies JE, Mohn WW, Eltis LD (2006) The complete genome of *Rhodococcus* sp. RHA1 provides insights into a catabolic powerhouse. Proc Natl Acad Sci USA 103:15582–15587

Meinwald J, Labana SS, Chadha MSJ (1963) Peracid reactions. III. The oxidation of bicyclo[2.2.1]heptadiene. J Am Chem Soc 85:582–585

Miyamoto K, Okuro K, Ohta H (2007) Substrate specificity and reaction mechanism of recombinant styrene oxide isomerase from *Pseudomonas putida* S12. Tetrahedron Lett 48:3255–3257

Montersino S, Tischler D, Gassner GT, van Berkel WJH (2011) Catalytic and structural features of flavoprotein hydroxylases and epoxidases. Adv Synth Catal 353:2301–2319

Mooney A, Ward PG, O'Connor KE (2006) Microbial degradation of styrene: biochemistry, molecular genetics, and perspectives for biotechnological applications. Appl Microbiol Biotechnol 72:1–10

Mor R, Sivan A (2008) Biofilm formation and partial biodegradation of polystyrene by the actinomycete *Rhodococcus ruber*: biodegradation of polystyrene. Biodegradation 19:851–858

Morrison E, Kantz A, Gassner GT, Sazinsky MH (2013) Structure and mechanism of styrene monooxygenase reductase: new insight into the FAD-transfer reaction. Biochemistry 52:6063–6075

Nikodinovic-Runic J, Flanagan M, Hume AR, Cagney G, O'Connor KE (2009) Analysis of the *Pseudomonas putida* CA-3 proteome during growth on styrene under nitrogen-limiting and non-limiting conditions. Microbiology 155:3348–3361

O'Leary ND, Duetz WA, Dobson ADW, O'Connor KE (2002a) Induction and repression of the *sty* operon in *Pseudomonas putida* CA-3 during growth on phenylacetic acid under organic and inorganic nutrient-limiting continuous culture conditions. FEMS Microbiol Lett 208:263–268

O'Leary ND, O'Connor KE, Dobson ADW (2002b) Biochemistry, genetics and physiology of microbial styrene degradation. FEMS Microbiol Rev 26:403–417

O'Connor K, Buckley CM, Hartmans S, Dobson AD (1995) Possible regulatory role for nonaromatic carbon sources in styrene degradation by *Pseudomonas putida* CA-3. Appl Environ Microbiol 61:544–548

O'Connor KE, Dobson AD, Hartmans S (1997) Indigo formation by microorganisms expressing styrene monooxygenase activity. Appl Environ Microbiol 63:4287–4291

Oelschlägel M, Gröning JAD, Tischler D, Kaschabek SR, Schlömann M (2012a) Styrene oxide isomerase of *Rhodococcus opacus* 1CP, a highly stable and considerably active enzyme. Appl Environ Microbiol 78:4330–4337

Oelschlägel M, Tischler D, Gröning JAD, Kaschabek SR, Schlömann M (2012b) Process for the enzymatic synthesis of aromatic aldehydes or ketones. Patent: DE 102011006459 A1 20121004

Oelschlägel M, Riedel A, Zniszczoł A, Szymańska K, Jarzębski AB, Schlömann M, Tischler D (2014a) Immobilization of an integral membrane protein for biotechnological phenylacetaldehyde production. J Biotechnol 174:7–13

Oelschlägel M, Zimmerling J, Schlömann M, Tischler D (2014b) Styrene oxide isomerase of *Sphingopyxis* sp. Kp5.2. Microbiology (UK) 160:2481–2491

Oelschlägel M, Zimmerling J, Tischler D, Schlömann M (2014c) Method for biocatalytic synthesis of substituted or unsubstituted phenylacetic acids and ketones having enzymes of microbial styrene degradation. Patent: DE 102013211075 A1 20141218; WO 2014198871 A2 20141218

Oelschlägel M, Heiland C, Schlömann M, Tischler D (2015a) Production of a recombinant membrane protein in an *Escherichia coli* strain for the whole cell biosynthesis of phenylacetic acids. Biotechnol Rep 7:38–43

Oelschlägel M, Kaschabek SR, Zimmerling J, Schlömann M, Tischler D (2015b) Co-metabolic formation of substituted phenylacetic acids by styrene-degrading bacteria. Biotechnol Rep 6:20–26

O'Leary ND, O'Connor KE, Duetz W, Dobson ADW (2001) Transcriptional regulation of styrene degradation in *Pseudomonas putida* CA-3. Microbiology 147:973–979

Otto K, Hofstetter K, Roethlisberger M, Witholt B, Schmid A (2004) Biochemical characterization of StyAB from *Pseudomonas* sp. strain VLB120 as a two-component flavin-diffusible monooxygenase. J Bacteriol 186:5292–5302

Panke S, Witholt B, Schmid A, Wubbolts MG (1998) Towards a biocatalyst for (*S*)-styrene oxide production: characterization of the styrene degradation pathway of *Pseudomonas* sp. strain VLB120. Appl Environ Microbiol 64:2032–2043

Park MS, Bae JW, Han JH, Lee EY, Lee S-G, Park S (2006) Characterization of styrene catabolic genes of *Pseudomonas putida* SN1 and construction of a recombinant *Escherichia coli* containing styrene monooxygenase gene for the production of (*S*)-styrene oxide. J Microbiol Biotechnol 16:1032–1040

Patrauchan MA, Florizone C, Eapen S, Gómez-Gil L, Sethuraman B, Fukuda M, Davies J, Mohn WW, Eltis LD (2008) Roles of ring-hydroxylating dioxygenases in styrene and benzene catabolism in *Rhodococcus jostii* RHA1. J Bacteriol 190:37–47

Paul CE, Tischler D, Riedel A, Heine T, Itoh N, Hollmann F (2015) Nonenzymatic regeneration of styrene monooxygenase for catalysis. ACS Catal 5:2961–2965

Pievo R, Gullotti M, Monzani E, Casella L (2008) Tyrosinase catalyzes asymmetric sulfoxidation. Biochemistry 47:3493–3498

Qaed AA, Lin H, Tang D-F, Wu Z-L (2011) Rational design of styrene monooxygenase mutants with altered substrate preference. Biotechnol Lett 33:611–616

Rao AVR, Gurjar MK, Kaiwar V (1992) Enantioselective catalytic reductions of ketones with new four membered oxazaborolidines: application to (*S*)-Tetramisole. Tetrahedron Asymmetry 3:859–862

Riedel A, Heine T, Westphal AH, Conrad C, van Berkel WJH, Tischler D (2015) Catalytic and hydrodynamic properties of styrene monooxygenases from *Rhodococcus opacus* 1CP are modulated by cofactor binding. AMB Express 5:30

Rioz-Martínez A, de Gonzalo G, Torres Pazmiño DE, Fraaije MW, Gotor V (2010) Enzymatic synthesis of novel chiral sulfoxides employing Baeyer-Villiger monooxygenases. Eur J Org Chem 33:6409–6416

Rioz-Martínez A, Kopacz M, de Gonzalo G, Torres Pazmiño DE, Gotor V, Fraaije MW (2011) Exploring the biocatalytic scope of a bacterial flavin-containing monooxygenase. Org Biomol Chem 9:1337–1341

Schmid A, Dordick JS, Hauer B, Kiener A, Wubbolts M, Witholt B (2001) Industrial biocatalysis today and tomorrow. Nature 409:258–268

Schröter F (2006) Biologische Abluftbehandlung der Pianofortefabrik Schimmel in Braunschweig. VDI (Verein Deutscher Ingenieure) AK Umwelttechnik, Braunschweig, Germany. Available at http://ut.vdi-bs.de/Schimmel.html. Accessed at 2011

Schulze B, Wubbolts MG (1999) Biocatalysis for industrial production of fine chemicals. Curr Opin Biotechnol 10:609–615

Sheldon RA (2007) Enzyme immobilization: the quest for optimum performance. Adv Synth Catal 349:1289–1307

Steinkellner G, Gruber CC, Pavkov-Keller T, Binter A, Steiner K, Winkler C, Lyskowski A, Schwamberger O, Oberer M, Schwab H, Faber K, Macheroux P, Gruber K (2014) Identification of promiscuous ene-reductase activity by mining structural databases using active site constellations. Nat Commun 5. doi:10.1038/ncomms5150

Teufel R, Mascaraque V, Ismail W, Voss M, Perera J, Eisenreich W, Haehnel W, Fuchs G (2010) Bacterial phenylalanine and phenylacetate catabolic pathway revealed. Proc Natl Acad Sci USA 107:14390–14395

Tischler D, Kaschabek SR (2012) In: Singh SN (ed) Microbial degradation of xenobiotics. Springer, Berlin, pp 67–99

Tischler D, Eulberg D, Lakner S, Kaschabek SR, van Berkel WJH, Schlömann M (2009) Identification of a novel self-sufficient styrene monooxygenase from *Rhodococcus opacus* 1CP. J Bacteriol 191:4996–5009

Tischler D, Kermer R, Gröning JAD, Kaschabek SR, van Berkel WJH, Schlömann M (2010) StyA1 and StyA2B from *Rhodococcus opacus* 1CP: a multifunctional styrene monooxygenase system. J Bacteriol 192:5220–5227

Tischler D, Kaschabek SR, Gassner GT (2011) StyA1/StyA2B, a unique flavin monooxygenase system. Flavins and flavoproteins, proceedings of the seventeenth international symposium, Berkeley, pp 307–312

Tischler D, Gröning JAD, Kaschabek SR, Schlömann M (2012) One-component styrene monooxygenases: an evolutionary view on a rare class of flavoproteins. Appl Biochem Biotechnol 167:931–944

Tischler D, Schlömann M, van Berkel WJH, Gassner GT (2013) FAD C(4a)-hydroxide stabilized in a naturally fused styrene monooxygenase. FEBS Lett 587:3848–3852

Toda H, Itoh N (2012) Isolation and characterization of styrene metabolism genes from styrene-assimilating soil bacteria *Rhodococcus* sp. ST-5 and ST-10. J Biosci Bioeng 113:12–19

Toda H, Imae R, Komio T, Itoh N (2012) Expression and characterization of styrene monooxygenases of *Rhodococcus* sp. ST-5 and ST-10 for synthesizing enantiopure (*S*)-epoxides. Appl Microbiol Biotechnol 96:407–418

Torres Pazmiño DE, Winkler M, Glieder A, Fraaije MW (2010) Monooxygenases as biocatalysts: classification, mechanistic aspects and biotechnological applications. J Biotechnol 146:9–24

Tuynman A, Vink MKS, Dekker HL, Schoemaker HE, Wever R (1998) The sulphoxidation of thioanisole catalysed by lactoperoxidase and *Coprinus cinereus* peroxidase: evidence for an oxygen-rebound mechanism. Eur J Biochem 258:906–913

Tuynman A, Schoemaker HE, Wever R (2000) Enantioselective sulfoxidations catalyzed by horseradish peroxidase, manganese peroxidase, and myeloperoxidase. Monatsh Chem 131:687–695

Ukaegbu UE, Kantz A, Beaton M, Gassner GT, Rosenzweig AC (2010) Structure and ligand binding properties of the epoxidase component of styrene monooxygenase. Biochemistry 49:1678–1688

Utkin I, Yakimov M, Matveeva L, Kozlyak E, Rogozhin I, Solomon Z, Bez-borodov A (1991) Degradation of styrene and ethylbenzene by *Pseudomonas* species Y2. FEMS Microbiol Lett 77:237–242

van Berkel WJH, Kamerbeek NM, Fraaije MW (2006) Flavoprotein monooxygenases, a diverse class of oxidative biocatalysts. J Biotechnol 124:670–689

van Hellemond EW, Janssen DB, Fraaije MW (2007) Discovery of a novel styrene monooxygenase originating from the metagenome. Appl Environ Microbiol 73:5832–5839

Velasco A, Alonso S, Garcia JL, Perera J, Diaz E (1998) Genetic and functional analysis of the styrene catabolic cluster of *Pseudomonas* sp. strain Y2. J Bacteriol 180:1063–1071

Volmer J, Neumann C, Bühler B, Schmid A (2014) Engineering of *Pseudomonas taiwanensis* VLB120 for constitutive solvent tolerance and increased specific styrene epoxidation activity. Appl Environ Microbiol 80:6539–6548

Wang J-C, Sakakibara M, Matsuda M, Itoh N (1999a) Site-directed mutagenesis of two zinc-binding centers of the NADH-dependent phenylacetaldehyde reductase from styrene-assimilating *Corynebacterium* sp. strain ST-10. Biosci Biotechnol Biochem 63:2216–2218

Wang J-C, Sakakibara M, Liu J-Q, Dairi T, Itoh N (1999b) Cloning, sequence analysis, and expression in *Escherichia coli* of the gene encoding phenylacetaldehyde reductase from styrene-assimilating *Corynebacterium* sp. strain ST-10. Appl Microbiol Biotechnol 52:386–392

Ward PG, de Roo G, O'Connor KE (2005) Accumulation of polyhydroxyalkanoate from styrene and phenylacetic acid by *Pseudomonas putida* CA-3. Appl Environ Microbiol 71:2046–2052

Ward PG, Goff M, Donner M, Kaminsky W, O'Connor KE (2006) A two step chemo-biotechnological conversion of polystyrene to a biodegradable thermoplastic. Environ Sci Technol 40:2433–2437

Warhurst AM, Fewson CA (1994) A review. Microbial metabolism and biotransformation of styrene. J Appl Bacteriol 77:597–606

Warhurst AM, Clarke KF, Hill RA, Holt RA, Fewson CA (1994) Metabolism of styrene by
   Rhodococcus rhodochrous NCIMB 13259. Appl Environ Microbiol 60:1137–1145
Zhang J-D, Li A-T, Yang Y, Xu J-H (2010) Sequence analysis and heterologous expression of a
   new cytochrome P450 monooxygenase from *Rhodococcus* sp. for asymmetric sulfoxidation.
   Appl Microbiol Biotechnol 85:615–624